Post-Industrial Urban Greenspace

Post-industrial urban spaces typically include abandoned factories, disused rail lines, old pits and quarries, and de-commissioned landfills. In these places, different visions compete for dominance with respect to current and future land uses. Neighbours often view such urban greenspace as polluted, unkempt and weedy, harbouring undesirable biophysical features and people. These are spaces that often become the focus of some form of revitalisation, reinvestment and restoration. From the perspective of civic authorities and urban planners, transforming post-industrial landscapes into disciplined and tended greenspace creates the urban conditions and signals of popular contemporary taste that attract investors, gentrifiers, and tourists. But post-industrial spaces are also places where unique and unpredictable human and ecological associations can emerge spontaneously. Such places may contain considerable ecological integrity and biodiversity and host human populations who find a home and respite in such ecologies. They also tell stories of an industrial and urban past that should be acknowledged, understood and (if suitable) celebrated. This volume explores the environmental justice and injustice dimensions of emerging urban post-industrial landscapes, including the ecological politics, cultural representations and aesthetics of these spaces.

This book was originally published as a special issue of *Local Environment*.

Jennifer Foster is an Associate Professor in the Faculty of Environmental Studies and Coordinator of the Urban Ecologies program at York University, Toronto, Canada. She is a Registered Professional Planner (RPP) with the Canadian Institute of Planners. Recent projects include the contemporary art exhibition *Land\Slide: Possible Futures* and the critical urban ecology study *From Rubble to Refuge*.

L. Anders Sandberg is a Professor and former Associate Dean in the Faculty of Environmental Studies at York University, Toronto, Canada. His two most recent books are *The Oak Ridges Moraine Battles: Development, Sprawl and Nature Conservation in the Toronto Region* (2013) and *Urban Trees, Forests, and Greenspace: A Political Ecology Perspective* (2014).

Post-Industrial Urban Greenspace

An environmental justice perspective

Edited by
Jennifer Foster and L. Anders Sandberg

Routledge
Taylor & Francis Group
LONDON AND NEW YORK

First published 2016 by Routledge

2 Park Square, Milton Park, Abingdon, Oxon OX14 4RN
711 Third Avenue, New York, NY 10017, USA

Routledge is an imprint of the Taylor & Francis Group, an informa business

First issued in paperback 2017

British Library Cataloguing in Publication Data
A catalogue record for this book is available from the British Library

ISBN 13: 978-1-138-91301-1 (hbk)
ISBN 13: 978-1-138-08569-5 (pbk)

Typeset in Times New Roman
by RefineCatch Limited, Bungay, Suffolk

Publisher's Note
The publisher accepts responsibility for any inconsistencies that may have
arisen during the conversion of this book from journal articles to book chapters,
namely the possible inclusion of journal terminology.

Disclaimer
Every effort has been made to contact copyright holders for their permission to
reprint material in this book. The publishers would be grateful to hear from any
copyright holder who is not here acknowledged and will undertake to rectify
any errors or omissions in future editions of this book.

Contents

Citation Information

The chapters in this book were originally published in *Local Environment*, volume 19, issue 10 (November 2014). When citing this material, please use the original page numbering for each article, as follows:

Chapter 1
Editorial: Post-industrial urban greenspace: justice, quality of life and environmental aesthetics in rapidly changing urban environments
Jennifer Foster and L. Anders Sandberg
Local Environment, volume 19, issue 10 (November 2014) pp. 1043–1048

Chapter 2
The greening of urban post-industrial landscapes: past practices and emerging trends
Christopher De Sousa
Local Environment, volume 19, issue 10 (November 2014) pp. 1049–1067

Chapter 3
Environmental gentrification in a post-industrial landscape: the case of the Limhamn quarry, Malmö, Sweden
L. Anders Sandberg
Local Environment, volume 19, issue 10 (November 2014) pp. 1068–1085

Chapter 4
Buried localities: archaeological exploration of a Toronto dump and wilderness refuge
Heidy Schopf and Jennifer Foster
Local Environment, volume 19, issue 10 (November 2014) pp. 1086–1109

Chapter 5
Re-presenting transgressive ecologies: post-industrial sites as contested terrains
Joern Langhorst
Local Environment, volume 19, issue 10 (November 2014) pp. 1110–1133

Please direct any queries you may have about the citations to
clsuk.permissions@cengage.com

INTRODUCTION

Post-industrial urban greenspace: justice, quality of life and environmental aesthetics in rapidly changing urban environments

One of the most fascinating features of urban post-industrial ecologies is the opportunity they present for novel assemblage. As the social, cultural, economic and political merge in expired, disused and/or abandoned industrial infrastructure, they reconstitute into new successional forms and processes. This is a remarkable interface through which urban nature is formed. It most often evolves rapidly, thanks to human neglect and disregard, through spontaneous, uncultivated progression into new socio-ecologies, and then frequently becomes the subject of "rediscovery" and redevelopment. The new greenspaces that are produced, whether unintended or by design, are also a critical interface for the enactment of environmental justice and injustice.

The original special issue was inspired by three sessions of the annual meeting of the Association of American Geographers (AAG) in 2012, in New York City, where fourteen presentations focused on the ecological formations and politics of post-industrial urban space. These were rich sessions that identified emerging dynamics and perspectives on post-industrial greenspace in a global context, focusing on both distributive and procedural environmental justice concerns. "Rediscovery" of urban space presents ample challenge to advancing justice concerns. But the AAG sessions also revealed that when former industrial spaces turn green, they not only induce outstanding and often unexpected ecological prospects, but can also function as frontiers for community rights to secure a self-determined quality of life.

The right to urban greenspace

While early developments in the field of environmental justice that focus primarily on the distribution of noxious materials remain foundational to exploration of post-industrial urban greenspace, much insight is drawn from recent contributions to the field that invests in a broader range of issues that include urban health, the Right to the City, accessible transit, affordable quality housing, economic opportunities and wealth creation related to the "green economy", and local climate mitigation (Anguelovski 2013), as well as the ecological politics of new noxious constellations, such as e-waste (Pellow 2006). Moreover, concern for racialised people and low-income neighbourhoods is increasingly joined with analytical attention to structural processes of capital accumulation, deindustrialisation and gentrification. Much of the work uncovered in the AAG sessions confirms Stein's (1998) and High and Lewis' (2007) suggestion that green sentiments have numbed the American middle-class' concern over urban industrial decline, and desensitised them to the significance of industrial physical ruins, toxic legacies and the deep painful history of working people who lived and worked in these environments.

Within this context, greenspace is increasingly being approached as a "right" of urban living, not only in terms of minimum standards for distribution, but also the quality of greenspace and targeting communities where it is most needed (Walker 2012). In these respects, as post-industrial environmental "bads" are stabilised and rendered into greenspace "goods", a wide range of justice concerns emerge. Are these new greenspaces actually improving the urban experience of marginalised people, and if so how and for whom? Do they respect the industrial legacies and lives that they oft times displace? Do they provide opportunities for the development of spectacular places that attract attention in a world where cities increasingly compete for capital investments, skilled workers and tourists? Do they establish the conditions for gentrification, resurfacing signals of blight and contamination with vibrant green, the allure of sustainable adaptive reuse and a trendy post-industrial patina? Do these transformations reinforce (or even introduce new) forms of material and procedural exclusion, and/or are they capable of producing more just cultural politics of representation? These questions form central themes that help advance an understanding of the complexity of transforming post-industrial urban space into green terrain surfaced through the AAG sessions, and this collection of papers.

Greening urban industrial scars

The ecological authorship of historical legacies is a notable concern for environmental justice. Research here demonstrates the varied ways that nature and ecological assemblages function as "politics by other means" in articulating the circumstances, conditions and experiences that produce urban environments. Memory of industrial past is typically selectively invoked, and planned and designed greenspaces rarely reflect the scars of industrialisation, the health hazards associated with contamination or the displacement of livelihoods associated with industrial decline. In this respect, the staging of greenspace is overtly political in determining who may rightfully inhabit rediscovered urban space, and the lapses of time where these spaces escape popular attention (following disuse and preceding redevelopment) offer some of the most incisive accounts of vulnerable lives on the urban margins.

Urban preoccupations, ideals and displacements

As new potential spaces of capital and economic activity, the interface between public and private realms is also a critical juncture for negotiating environmental justice, particularly in terms of determining whose ideals and preferences inform decisions, how, and where these fit into civic preoccupations and global circuits of capital. The risk of gentrification was an immediate concern for almost all of the work of the AAG sessions and the papers in this special issue. Determining where, how much, and which nature to either disregard and allow to take its own course or remove and refabricate through planning and design processes is a central determinant of gentrification pathways. The specific conceptions of nature that are invoked, the types of access to nature and the extent to which local communities shape decision-making processes have a huge influence on the risk of displacement. It is now well established that brownfields are typically located in poor and minority communities (Ringquist 2005, Campbell *et al.* 2010), the pace of clean-up is the slowest in minority communities (Eckerd and Keeler 2012) and urban redevelopment that brings about clean-up tends to favour marketable sites in attractive locations over locales that experience "place stigmatization" (McCarthy 2009, Pearsall 2013). Meanwhile, the outcomes of decontamination and redevelopment of brownfields are not always just, often prompting an increased cost of living and vulnerability to contraction of social housing, raising deep concerns about how different communities experience urban sustainability initiatives such as conversion of "vacant" post-industrial land into greenspace (Essoka 2010, Pearsall 2010).

The political work of aesthetics and authenticity

One of the most under-explored themes of environmental justice features prominently in this collection of work: the socio-politics of environmental aesthetics. If ecology functions as both a refraction and agent of cultural change, emergent aesthetic forms play a critical role in determining the justice dimensions of post-industrial urban greenspace (Edensor 2005). As the ruins of industrial space are increasingly fetishised in urban design and Western consumer desires, narratives of "authenticity" are articulated not only by literal means, but also through integration of ecological qualities (such as native ecosystem types and species associations) that appeal to contemporary interpretations of urban sustainability but do not necessarily guarantee or even advance social justice or enhanced ecological quality. The aesthetic dimensions of post-industrial greenspaces – how they look, feel and the socio-cultural signifiers they produce and mobilise – are a strong foundation for the political work of these spaces. In this sense, ecological systems perform and bolster notions of authenticity, legitimising adaptations to former industrial sites with ecological accounts of belonging and consistency with previous circumstances. Yet there are also examples that promote aesthetic formations that run counter to the mainstream, marketable, cleaned up aesthetic by seeking to reveal their sordid and working pasts, as well as their connections to the creation of new production facilities elsewhere, often in third world locations with poor working conditions and environmental records (Edensor 2005, Fassi 2010), or by allowing *terrain vague* and other "hands off" aesthetic conceptions to flourish (Kamvasinou 2006, Gandy 2013, Foster 2014). The selection of which industrial artefacts, which ecologies and which stories to express through the aesthetics of greenspaces that emerge in the wake of industrial decline are politically poignant, situating these sites within the cities and networks within which they are embedded, but also with the people whose livelihoods intersect historically and in the present.

Greenspace through development, division, design and erasure

In the following, we present four papers that tackle the above themes. To begin with, De Souza provides a policy perspective based on pluralist assumptions of post-industrial landscapes. He describes urban greening as a positive development that provides the urban population with a respite from the everyday existence in the built environment and the potential role of former industrial sites or brownfields in this quest. He draws on extensive research on post-industrial landscapes in the USA and Canada to provide a record on the initiators of the greening of such landscapes, their characteristics, and the costs and barriers associated with their rehabilitation. He then explores three brownfield case studies: Elmhurst Park in New York City, the South Waterfront in Portland and the Menomonee Valley in Milwaukee. The rehabilitation of all areas, De Souza suggests, is the result of national and local actors taking on the role of raising money, securing land and overcoming the local pollution to establish green areas that now serve the urban population at large.

The three following contributions build on De Souza's insights by asking questions about the social divisions and tensions and human–non-human interactions that surround post-industrial landscapes. These accounts rest on a broad politics informed by social and economic power as well as history and aesthetics of the non-human in influencing the shape of post-industrial landscapes.

Sandberg provides a case study of a phased out limestone quarry, the Limhamn Kalkbrott in Malmö, Sweden, and its current position as an internationally recognised nature preserve that contains several endangered and at-risk species. He argues that the greening of Kalkbrottet is part of two processes, one exclusive and exclusionary, the other democratic and inclusionary. On the one hand, Kalkbrottet is part of a process of neo-liberalisation and gentrification of Malmö, a situation where the scenic properties of the quarry are part of the City's efforts to attract capital and skilled and educated workers. The high-priced condominiums and rental housing developments that now surround Kalk-brottet, Malmö's Grand Canyon, are part of that process. On the other hand, Kalkbrottet may also be a phenomenon that resists the neoliberalisation of the city. This is reflected in the roles played by rare flora and fauna that have unexpectedly appeared in the quarry and the strong voice of city ecologists in support of the designation of the quarry as a nature preserve. Sandberg also argues that the historical legacy of the quarry and its current use by legitimate and illegitimate visitors may have the potential to make Kalkbrottet a focal point for a more socially just and environmentally sustainable Malmö.

Schopf and Foster investigate a constructed post-industrial landscape at Leslie Street Spit in Toronto, Canada, which is world renowned as a bird sanctuary and a spontaneously emerging ecological reserve that is now a major recreational area for Torontonians. The Spit, jutting five kilometres into Lake Ontario, is composed of construction debris from the process of tearing down and re-building the city. Using an archaeological method, Schopf and Foster show that the Leslie Street Spit is composed of much more than the "clean fill" claimed by the City of Toronto. In fact, the Spit articulates the creative destruction of the city through the demolition of neighbourhoods that hindered the city's advance into modernity. The rubble from one part of the Spit, for example, contains intact household items from the working class districts that were destroyed in the 1960s to make way for the concrete high-rise buildings attracting capital at the time. At another site, the authors expose a beach of bricks which were formerly the building blocks of single-family houses that comprised a community which the city allowed to be transformed into office towers in the 1980s. The material of the Spit, Schopf and Foster argue provides critical memory of lost and largely forgotten neighbourhoods that are part of the urban fabric, and can inspire new ways of thinking about, living in and re-imagining the city.

Langhorst finally uses the examples of Landschaftspark Duisburg-Nord (in Duisburg, Germany) and the High Line (in New York City) as examples of different aesthetics or representational agencies of ecology in the context of post-industrial greenspaces. Duisburg-Nord of the Ruhr region is an expired steel and coal complex that is now an extensive green and recreational area with many of the old industrial buildings and structures intact, while High Line is a decommissioned elevated freight train line in Manhattan that has been transformed into a celebrated green area with a pedestrian walkway. Langhorst argues that Duisburg-Nord represents an aesthetic that favours the spontaneous interaction of humans and non-human nature, where plants and flowers are free to invade the old industrial buildings, and where people's activities are largely unrestricted. At the High Line, by contrast, another nature aesthetic prevails, one that controls human and non-human activities.

These essays, taken together, examine a variety of post-industrial spaces and their prominent yet different roles in the greening of the city. They also reflect different ways of thinking about such landscapes. De Souza shows how the greening of industrial sites has become a common development quest, often hard fought for, by city governments but typically appreciated by the broader urban population. Sandberg points to the presence of post-industrial greenspaces that are contested by proponents and opponents of the neoliberal and gentrified city. Schopf and Foster propose that the material foundation of post-industrial landscapes may be composed of lost stories that are central to environmental justice concerns relating to the lives of real people, industries and urban development trajectories. Langhorst, finally, stresses that the representation of the aesthetics of the non-human is an important agent in shaping the character of the post-industrial landscape.

The politics of post-industrial "rediscovery" and greenspace creation

The greening of post-industrial landscapes is not only a matter of the creation of an unquestionable and unqualified public good but also a process infused with political and environmental justice concerns. Clearly, post-industrial sites are contested places. On the one hand, they are subject to appropriation by the privileged and wealthy and the process of global capital accumulation. They also often become part of gentrified communities enamoured by a post-industrial chic couched in a variety of green aesthetics and socio-political contexts. Sometimes these green contexts are highly regulated, giving them a disciplined character that corresponds to the aspirationally safe, ordered and exclusive built environment of which they are a part. On other occasions, some allowance is made for non-human spontaneity, typically only when it does not threaten gentrified communities. Yet post-industrial sites may also challenge dominant social class and capital accumulation processes. This special issue on post-industrial urban greenspace speaks to and raises questions about this concern. These papers suggest a range of complex yet pressing questions at the forefront of urban environmental justice. What are the potential of the marginalised and racialised to seize on and rally the post-industrial green to transform and make more livable their neighbourhoods? What can the non-human and alternative environmental aesthetics contribute to such an endeavour? And what types of alliances can be built between nature advocates, the non-human that finds a place in the post-industrial landscape, and residents and former workers who occupy the post-industrial?

Jennifer Foster
L. Anders Sandberg

References

Anguelovski, I., 2013. New directions in urban environmental justice: rebuilding community, addressing trauma, and remaking place. *Journal of Planning Education and Research*, 33 (2), 160–175.

Campbell, H., Peck, L., and Tschudi, M., 2010. Justice for all? A causal analysis of toxics release inventory facility location. *Review of Policy Research*, 27 (1), 1–25.

Eckerd, A., and Keeler, A.G., 2012. Going green together? Brownfield remediation and environmental justice. *Policy Sciences*, 45 (4), 293–314.

Edensor, T., 2005. *Industrial ruins: space, aesthetics, and materiality*. Oxford: Berg.

Essoka, J.D., 2010. The gentrifying effects of brownfield redevelopment. *Western Journal of Black Studies*, 34 (3), 299–315.

Fassi, A., 2010. Industrial ruins, urban exploring, and the postindustrial picturesque. *The New Centennial Review*, 10 (1), 141–152.

Foster, J., 2014. Hiding in plain view: vacancy and prospect in Paris' Petite Ceinture. *Cities*, 40 (B), 124–132.

Gandy, M., 2013. Entropy by design: Gilles Clément, Parc Henri Matisse and the limits to avant-garde urbanism. *International Journal of Urban and Regional Research*, 37 (1), 259–278.

High, S., and Lewis, D., 2007. *Corporate Wasteland: the landscape of memory and deindustrialization*. Ithaca, NY: Cornell University Press.

Kamvasinou, K., 2006. Vague parks: the politics of late twentieth-century urban landscapes. *Arq*, 10 (3/4), 255–262.

McCarthy, L., 2009. Off the mark? Efficiency in targeting the most marketable sites rather than equity in public assistance for brownfield redevelopment. *Economic Development Quarterly*, 23 (3), 211–228.

Pearsall, H., 2010. From brown to green? Assessing social vulnerability to environmental gentrification in New York City. *Environment and Planning C: Government and Policy*, 28 (5), 872–886.

Pearsall, H., 2013. Superfund me: a study of resistance to gentrification in New York City. *Urban Studies*, 50 (11), 2293–2310.

Pellow, D.N., 2006. Transnational alliances and global politics: new geographies of urban environmental justice struggles. *In*: N. Heynen, M. Kaika, and E. Swyngedouw, eds. *In the nature of cities: urban political ecology and the politics of urban metabolism*. London: Routledge, 226–244.

Ringquist, E., 2005. Assessing evidence of environmental inequities: a meta-analysis. *Journal of Policy Analysis and Management*, 24 (2), 223–247.

Stein, J., 1998. *Running steel, running America: race, economic policy and decline of liberalism*. Chapel Hill: University of North Carolina Press.

Walker, G., 2012. *Environmental justice: concepts, evidence and politics*. London: Routledge.

The greening of urban post-industrial landscapes: past practices and emerging trends

Christopher De Sousa

School of Urban and Regional Planning, Faculty of Community Services, Ryerson University, Toronto, Ontario, Canada

Public, private, and non-profit entities are increasingly engaged in greening post-industrial landscapes in an effort to achieve a broad array of aesthetic, infrastructure, recreational, ecological, and economic development objectives at various scales. Despite this growing level of interest, however, these projects continue to face numerous challenges related to financing, land acquisition, soil contamination, and concern regarding long-term maintenance, just to name a few. This paper begins with an overview of the "nature" of greening activity that has taken place in the USA and Canada and then focuses on three case studies – Elmhurst Park New York City, South Waterfront Portland, and Menomonee Valley Milwaukee – in order to illustrate the planning processes involved in their remediation and development. Key lessons are then drawn, with a particular emphasis on the growing need to attract buy-in and funding by linking greening with other forms of development and broader urban sustainability initiatives.

While the greening of post-industrial landscapes was considered a fringe issue by most planners and economic development officials less than a decade ago, the incorporation of ecological and recreational features into redevelopment projects has become almost common practice as cities try to achieve a broader range of environmental, economic, and social goals associated with sustainability. More and more frequently, public, private, and non-profit entities are involved in greening post-industrial landscapes in a strategic manner to achieve aesthetic, infrastructure, recreational, ecological, and economic development objectives at various scales (Bradshaw 2000, ICMA 2002, Burger *et al.* 2004, Byrne *et al.* 2007). These multiple objectives are not only being contemplated for urban park projects, but increasingly for sites being redeveloped for other uses as well.

This paper begins by providing an overview of the type of greening projects that have been developed on brownfields in the USA and Canada and the planning processes involved in their remediation and redevelopment. Three case studies are then examined to illustrate the planning and development process in more detail. The case studies include Elmhurst Park, a brownfield to green space project in New York City, the South

Waterfront mixed-use development in Portland, and the Menomonee Valley redevelopment in Milwaukee, which uses green infrastructure to convert a post-industrial district into a modern industrial district. The paper concludes by summarising the barriers associated with greening and offers lessons for overcoming them. It argues that while traditional barriers to greening still exist, tying it to broader urban and neighbourhood sustainability, restoration, and economic development initiatives is becoming more widely accepted, thus providing an opportunity to further transform the urban landscape and enhance urban ecologies.

What is urban greening?

Urban greening is understood generally to mean the creation of green spaces within a city's built-up environment, which includes the production of parks, public spaces, gardens, outdoor athletic facilities, natural habitats, greenways, and children's playgrounds through redevelopment (Garven and Berens 1997, Harnik 2000, Bunster-Ossa 2001). It can also refer to the preservation, protection, and enhancement of those natural areas within the urban fabric that have been, for some reason or other, left undisturbed (Schilling and Logan 2008, CL:AIRE 2009). While the term greening is now being used more widely to refer to the incorporation of environmental ideas into business activity, it is used here more conventionally, but loosely, to refer to the preservation and development of parks and green spaces in cities, whether this be the primary mandate of a project or a complement to another use.

The rapid shift from a primarily rural/agricultural to an urban/industrial economy in the early nineteenth century brought tremendous population growth to urban areas throughout North America. Coinciding with this growth were the pollution and health problems associated with the congested industrial city. Thus, the need for increased recreation, better public health conditions, and relief from the toils of industrial labour prompted initial calls for urban greening via the creation of parks in both the USA and Canada (Platt 2010).

The first notions of the urban park were to be realised by a culturally elite group of leaders and designers such as Andrew Jackson Downing, Frederick Law Olmsted, and Calvert Vaux. The grand central park model of New York would be replicated throughout North America in the second half of the nineteenth century. According to the typology put forward by Cranz (1982), which takes into account both social and design issues, the aim of parks has shifted and evolved, moving from the creation of elite cultural spaces to providing services for families, children, workers, and the general public. The "Reform Park" movement (1900–1930) focused on providing play and recreation opportunities in smaller, more accessible, parks distributed evenly throughout cities (Cranz 1982). In a later paper, Cranz and Boland (2004) note that this was followed by the Recreational Facility period (1930–1965), aimed at supplying active play facilities and services, and the Open Space System (1965–no end date provided), where trees, grass, and open areas allowed for more free-flowing play and expression. The authors add a fifth park model, the Sustainable Park (1995 to present), which indeed reflects many of the desires and designs of those currently engaged in the greening of post-industrial landscapes in urban areas. In it, native plants, permeable surfaces, green infrastructure, and resource self-sufficiency are used as key elements of park design to allow for a wide range of active and passive recreational activities as well as nature viewing, education, and stewardship (Cranz and Boland 2004). This type of green space is intended to benefit not only human users in various ways, but wildlife and the planet as well, which allows it to be easily woven into the broader discourse on urban sustainability.

Turning brownfields into green space

Research on the greening of post-industrial landscapes and other brownfields has expanded considerably over the last few years (Loures and Panagopoulos 2007, Schilling and Logan 2008, Siikamäki and Wernstedt 2008, Hofmann *et al.* 2012). This section summarises previous brownfields-to-green space research undertaken by the author in North American cities in an effort to paint a broader picture of the nature and scope of the projects being undertaken as well as to highlight the planning and development processes and challenges common to the 30 projects examined (De Sousa 2003, 2004, 2006, 2010).

The greening of brownfields in both the USA (De Sousa 2004) and Canada (De Sousa 2003) has led to the creation of an array of green spaces and made a significant contribution to urban park inventories. In aggregate, the 20 greening projects examined in a study of US cities generated approximately 1144 new acres of green space (463 hectares) that range in size from 2 to 400 acres (0.81–162 hectares). The mean size of the projects was 57 acres (23.1 hectares) and the median size 14 acres (5.7 hectares), with most (13) of the projects in the 2–20 acre range (0.81–8 hectares) (De Sousa 2004). The brownfields were developed into various types of open space, including linear parks or greenways (7), large-scale parks intended for multiple recreation uses (6), smaller spaces for neighbourhood recreation (4), and habitat restoration areas (3). Thirteen of the projects involved both recreation and habitat creation, and 11 included the creation of educational amenities (i.e. markers, signs, site tours, and visitors/education centres). Eight of the projects also involved the preservation of historically significant structures and areas celebrating the site's industrial past (e.g. buildings, bulkheads, rail cars, and machinery).

Smaller neighbourhood projects, like the Elmhurst Park case described later in this paper, have been developed on brownfields in both countries to provide a widerange of active and passive recreation activities and are typically constructed in mid-density older neighbourhoods that lack parks (De Sousa 2003, 2004). The decision to construct such parks is often reactive, with governments under severe pressure from residents to increase park space and oppose other forms of development. The most prevalent type of post-industrial greening in both of the countries studied has been the construction of greenways and green spaces connecting linear trails (De Sousa 2003, 2004, 2006). These projects, like the South Waterfront and Menomonee Valley cases described later in the paper, are either built on former railway lines or on industrial properties located along waterways that often have to be painstakingly sewn together by negotiating public access to multiple properties, acquiring others outright, and by connecting individual park developments. Interest in greenways is widespread because in addition to providing space for recreation, they provide an aesthetically pleasant transportation route, connect habitat and species migration corridors, and are usually a component of a broader urban redevelopment strategy that sees residential and commercial development sprout up along post-industrial waterfronts and riverfronts.

While many cities have been successful at converting post-industrial landscapes into an array of green spaces, the planning and implementation process is typically more complex than for other brownfield redevelopment projects and often requires the interaction of various levels of government, private-sector participants, non-profit organisations, and community-based groups. In the brownfield-to-park projects studied by the author, local governments were most often the ones taking the lead in coordination, while upper levels of government (e.g. federal and state in the USA and provincial and regional in Canada) also played an important role by providing land, funding, technical expertise, and some assistance with project coordination. Developers, landowners, and other private-sector partners were involved in slightly over half of the US projects, both directly (via site construction or property donation) and indirectly (via coordination), but the success of their own developments had to be tied to the new green space in some way.

Nonprofits are playing an increasingly active role in greening post-industrial districts, particularly in the USA. Some groups, such as the Trust for Public Land and Groundwork USA (and Evergreen in Canada), have broadly defined missions aimed at promoting the quality of urban life, while others such as the Rails-to-Trails Conservancy have more narrow missions associated with a particular type of greening. These national-level nonprofits, most of which have regional or local branches, have found themselves at the forefront of greening post-industrial areas in many respects (i.e. raising awareness, remediation, financing, planning and implementation, and post-development promotion). Their most notable role is often acting as a bridge between the public sector, the private sector, the local community, and even local nonprofits who may have initially requested their guidance. Many have become very experienced in fundraising and dealing with risk (whether it is legal, financial, pollution, or technological), and posses technical skills (fundraising, lobbying, design, cleanup techniques, real estate acquisition, etc.) that other stakeholders cannot provide.

Another issue standing in contrast to other kinds of profit-oriented brownfield redevelopment initiatives is the very important role communities play in greening (Eisen 1999). Pressure for greening and decisions related to all aspects of redevelopment projects are typically influenced by community involvement and input. Such involvement comes about through various structures and in diverse contexts, including, for example, consultation forums, design charrettes, working groups, committees, site visitations, and educational tours. Community participation often continues beyond project planning and development, with residents becoming involved in long-term management and maintenance of green spaces (via walking tours, educational/event programming, monitoring of habitat, etc.).

All in all, the diverse array of partners and their efforts to achieve broad economic, social, and environmental goals demonstrate that greening projects are particularly challenging, requiring collaboration among various, and often historically adversarial, stakeholders. For this reason, such projects take a long time to bring to fruition, requiring from 2 to 13 years to complete for the US projects examined (2004), and 3 to 5 years in research undertaken in Canada (2003). As with any brownfield project, greening also involves additional costs associated with site assessment and remediation. Information for 19 of the 30 projects examined in the USA and Canada reveals that the average cost for site assessment and remediation was approximately US$1.2 million a project or US$128,000 per acre, with costs for the US projects, as well as their duration, being slightly higher than those in Canada due to the redevelopment of more large-scale recreational facilities in heavily impacted industrial zones. It should be noted, however, that cleanup costs ranged extensively from US$40,000 to US$13 million per project. In the US cases, site preparation and remediation costs represented approximately 19% of the total project costs (based on estimates from 14 projects), but about 25% of total costs in the Canadian case.

In terms of cleanup, many of the projects (10) examined involved the use of capping, landscape features, or other engineered barrier methods to secure contaminants and minimise exposure. More traditional approaches such as off-site disposal (i.e. dig-and-dump) and pump-and-treat technologies were also used. Although they are becoming more prevalent, phytoremediation, bioremediation, and natural attenuation were only used in a few instances. Multiple methods were employed at 10 of the 11 US sites where information was available. Soil is often shifted throughout the site and managed depending on its contamination levels and risk. This makes it somewhat difficult to separate contaminated soil-management cost from construction cost because a cap can serve multiple uses (e.g. a road or parking lot, soil for grading and planting, a tennis court, etc.). It is important to note that brownfields converted into green space normally require a higher standard of cleanup than for commercial or industrial development because the level of exposure to soil is anticipated to be greater and those using the site will typically be younger.

The average total cost for a greening project in the US cities was US$6.05 million based on 18 projects (= US$3.99 million median cost or US$750,000 per acre/US$1.85 million per hectare). For those in Canada, it was CDN$1.8 million per hectare (CDN$660,000 per acre), while the median cost was CDN$580,000 per hectare (CDN$211,000 per acre). The discrepancy between the mean and the median values is no doubt due to the high costs tied to projects involving constructing or rehabilitating buildings/facilities for recreation-oriented purposes versus those involving less-costly ecological restoration or passive recreation. What is similar in all cities is that the government is responsible for covering virtually all the costs involved in greening. Unlike industrial, residential, or other types of brownfield redevelopment, where government funding is provided mainly to help with assessment and remediation costs, it is typically the only source of funding for assessing, remediating, and constructing greening projects. The city government most often pays for neighbourhood parks, while various levels of government sponsor many of the larger parks and may assist with assessment and remediation costs. That said, local governments are also responsible for, and concerned about, the long-term maintenance of these new spaces.

Despite the costs and other challenges, greening is increasingly seen as important not only as a reprieve from work and crowdedness, but for its many other environmental, economic, and social services. The case studies below provide a more in-depth look at projects that seek to harvest the broader potential of greening to achieve sustainability outcomes. Information gathering for these case studies was funded by the US Environmental Protection Agency (EPA) and obtained from available project reports, site visits, and structured face-to-face interviews that typically lasted one hour with four to six different stakeholders (including developers, planners, consultants, and community representatives). A questionnaire was used to gather information about site history, cleanup, project attributes, sustainability features, and planning and development procedures, as well as about the impressions of the interviewees regarding the barriers to implementation, how these were overcome, the outcomes that have ensued, and lessons learned.

Elmhurst Park, New York City

Spurred on by Mayor Bloomberg's comprehensive sustainability planning effort, PlaNYC, the City of New York has been extremely proactive in both the redevelopment and greening of its post-industrial landscapes (City of New York 2011, p. 57). Like many other municipal brownfields initiatives in the USA, the New York City Brownfield Cleanup Program, launched in 2010, helps coordinate the activities of different departments and agencies, and supports redevelopment activity. It is also the first programme in the country to manage the review of remediation plans on behalf of the state and issue a Notice of Completion to release developers from liability once cleanup is performed to standards. Furthermore, this innovative programme adds a Sustainability Statement to the Remedial Action Work Plan to acknowledge projects with sustainability features, such as green building, energy efficiency, or stormwater management. This new programme is intended to help the city react more quickly to assessment and remediation applications, and to manage its brownfields inventory in a more holistic and strategic manner by identifying spaces of opportunity for greening, public services, sustainable urban restoration, and other economic development activities. The City of New York Parks and Recreation Department's (2010) new landscape guidelines, *High Performance Guidelines: 21st Century Parks for NYC*, also raise the bar on sustainable park design, construction, and maintenance, and prioritise brownfields and recovered sites as major opportunities for urban greening.

Elmhurst Park is one of several new brownfields-to-park projects in New York (e.g. Fresh Kills, the High Line, Concrete Plant Park). The six-acre (2.5 hectare) neighbourhood park located in the borough of Queens replaces an industrial storage facility owned by KeySpan Energy Corporation that consisted of two above-ground gas tanks. Although the newest Elmhurst Gas Tanks were constructed in the 1970s, the site had been home to gas storage activities since 1910. Investment in new storage and distribution technologies prompted KeySpan to abandon the new tanks in the 1980s (Greene 2008). Until they were dismantled in 1996, the idle "white-and-red" gas tanks, both 275 feet in diameter, served as a local commuter landmark (Hughes 2006, Lopes 2011). The idle property stood in sharp contrast to the rapid growth of the surrounding lower-middle-class and ethnically diverse neighbourhood, which had become a popular destination for new immigrants during the 1980s and 1990s. While the private sector quickly responded to the influx of residents to the neighbourhood, public institutions were slower to provide the services and amenities needed, including green space (Hughes 2006). Indeed, the Trust for Public Land (2013) *ParkScore Index* (see Figure 1) notes that while Elmhurst Park (bottom left) has addressed park need issues within a quarter mile, many blocks of "very high" and "high" park need still lie to the north, south-east, and south-west of the site in some of the lower-income communities.

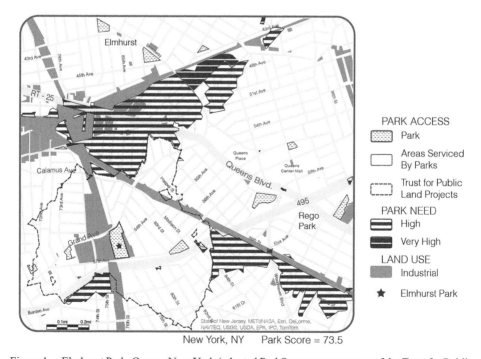

New York, NY   Park Score = 73.5

Figure 1.   Elmhurst Park, Queens New York (adapted ParkScore map courtesy of the Trust for Public Land).

In September 2001, KeySpan began remediating the property based on a New York State Department of Environmental Conservation (DEC) Voluntary Cleanup Agreement with the intention of selling it for redevelopment (Lopes 2011). After deconstructing the two above-ground storage tanks, remediation was completed to residential standards using a "grid method" that involved subdividing the entire area into blocks and then managing each one individually (Anon 2009, p. 1). The assessment revealed that the entire site was contaminated with lead, which necessitated the removal and off-site disposal of 18,000 tons of soil. In order to develop the property into a park, the site grade was raised by 6–12 feet using clean soil (New York State DEC 2012).

In August 2002, three preliminary plans were put forward for the site: (1) a residential development, (2) a 34-store mall and parking lot with a capacity for 1200 cars, and (3) a big box retail establishment (Juniper Park Civic Association 2003). Local residents, collaborating primarily through existing civic action groups, including the association for Juniper Park located about half a kilometre to the south, voiced their concern regarding the proposals and requested the donation of the land for use as a public park. KeySpan responded that donating the land valued at $13 million was not feasible according to regulations established by the Public Service Commission.

Between August 2002 and November 2003 several meetings between the civic organisation, KeySpan, private developers, elected officials, and local planning boards were held to negotiate the fate of the vacant property. In November 2002, private developers approached KeySpan with an offer to purchase the property and convert it into a big box development. Several months after the offer was tentatively accepted, in March 2003, elected officials requested delaying the sale of the property while they determined if municipal funds could be allocated to purchase the land. Although KeySpan agreed to the delay, it also proceeded to complete contract negotiations with the developer for a big box centre. In the meantime, local councillors successfully secured $3 million to facilitate public acquisition of the property (Juniper Park Civic Association 2003, p. 3).

Subsequent communications between stakeholders became increasingly hostile, and residents began organising a protest scheduled for 8 November 2003 at KeySpan's Brooklyn-based headquarters, and also "demand[ed] that elected officials pressure KeySpan" to abandon the deal (Juniper Park Civic Association 2003, p. 5). In October 2003, a sequence of meetings between elected officials escalated to the Mayor's office, where it was agreed that the City should buy the land. Mayor Bloomberg contacted the CEO of KeySpan to negotiate the acquisition. An announcement was made days before the scheduled protest that the "contract for building of a Home Depot would be deemed null and void and that the City intends to purchase this land for placement of a passive park" (Juniper Park Civic Association 2003, p. 6). KeySpan sold the Elmhurst Gas Tanks property to the City's Department of Parks and Recreation in 2005 for $1 (New York City DEC 2012).

Planning for the new park was completed in 2006 with input from local residents. The final project vision for the park included a playground, recreational pathways, public art, and landscaping based on the site's pre-industrial land characteristics. As noted by an interviewee from the city, the park plan was implemented in three phases between 2007 and 2012 to distribute the financial cost of the development over several budget cycles as funding, approximately $20 million, was provided through the Mayor's capital budget. Phasing also prioritised infrastructure projects that addressed site-specific considerations relating to its former brownfield status. The park's characteristics and development reflect an integrated approach that considers local heritage, interpretive design, and sustainability. Indeed, Elmhurst Park was also designed with reference to the Sustainable Sites

Initiative, which provides a framework for landscape design through a checklist that allocates points to project components that preserve and/or enhance ecosystem processes; and based on best practice performance benchmarks that are compatible with the US Green Building Council's Leadership in Energy and Environmental Design (LEED) green building rating system.

Phase I of the Elmhurst Park project focused primarily on preparing the site for further development. As per the soil-management plan, a representative from the city noted that 4000 cubic feet of clean fill material was used to cap the property and warning mats were laid below the clean fill material to alert prospective developers about the property's historic condition. A stormwater management system was also implemented to capture runoff from storm events on-site, and to prevent precipitation from draining into the city's sewer infrastructure. Precipitation is collected over approximately half the site in an underground retention system that consists of perforated pipes and broken stone wrapped in filter fabric (Lopes 2011). Stormwater is also collected in underground storage tanks and used to irrigate the park's landscaping. Phase I also implemented the park's site plan infrastructure through the construction of sidewalks, utilities, fencing, and landscaping (Lopes 2011). Given that the property had been denuded of vegetation, nearly 500 trees were planted to renew ecological functions, screen out surrounding land uses, commemorate former orchards, and create a park-like atmosphere. Phase I was completed in late 2008 at a cost of $5 million (Anon 2009, p. 1).

Phase II of the redevelopment focused on the construction of the park's playground and recreational amenities. The playground was designed in part to promote sustainability and energy literacy amongst children through experiential and interpretive features, with a nod to the park's industrial heritage. In addition to traditional playground equipment such as swings and slides, "children can pedal [one of three] stationary bike[s] to provide a light effect on a nearby column" (Lopes 2011, p. 2). A decorative fountain, benches, and lighting were also installed in the park. An existing building, previously occupied by KeySpan, was retrofitted for use by the park's maintenance staff. Additional landscaping as well as another 150 trees concluded the second phase of the park's development and the construction of a new ultra-modern and high-efficiency comfort station, completed in 2012, marked the final phase of Elmhurst Park's development.

Elmhurst Park is profiled as a best practice in greening post-industrial landscapes because of the integrated approach to park planning that emphasises sustainability. As conveyed by the Department of Parks and Recreation's Senior Project Manager for Sustainability, "there was no distinction between park design and sustainable park design". Despite the design innovations, the project provides a rather typical example of a small brownfield-to-park conversion in terms of the city's reactive response to neighbourhood advocacy, the cleanup approaches used, and the use of public resources for project development, notwithstanding the low-cost land donation from the private owner. What is changing for small greening projects like these is that they are being increasingly embraced by a broader array of public agencies and other stakeholders as an essential component of sustainable urban restoration, and not just a niche project for park departments. Indeed, the Parks Department received a Big Apple Brownfield Open Space Award for Elmhurst Park in 2009 from the NYC Brownfield Partnership, which is a non-profit organisation composed of partner organisations that include environmental consulting firms, environmental law organisations, environmental remediation contracting organisations, brownfield redevelopment organisations, and community-based organisations in New York City.

South Waterfront Portland

The South Waterfront District redevelopment project is the largest urban renewal initiative in Portland's history. The redevelopment is transforming a 140-acre (57 hectare), underutilised, and isolated industrial district into a mixed-use, transit-oriented neighbourhood while greening and reclaiming public access to 1.2 miles along the Willamette River. Of the three cases examined here, the South Waterfront redevelopment provides a typical, albeit large-scale, example of the efforts many North American cities are making to green brownfields along waterways (see for example, Minneapolis' Mill District, Philadelphia's Waterfront, and Vancouver's False Creek).

Industrial activities characterised the landscape of the South Waterfront for most of the twentieth century (Portland Bureau of Planning 2002, p. A-6). The Portland Lumber Company was one of the first businesses to establish in the district, supplying heat and electricity to Portland's downtown core beginning in the 1880s (The Center for Brownfields Initiatives 2003). The locational assets of the area appealed to businesses that relied on river frontage, road, railway access, and proximity to downtown. Other businesses that contributed to the industrial flavour of the area, while supporting its timber and shipping trades, included a metal fabrication plant, chemical manufacturers, an aluminium smelting operation, and various salvaging facilities. By 1910, these businesses had permanently transformed the riparian forest into an industrial hub. By the 1960s however, the momentum of the South Waterfront's industrial development began to decline as technological innovation in several sectors began to displace older industrial practices. The development of two freeways, Harbor Drive in the 1940s and the I-5 in the 1960s, compounded this decline by acting as physical barriers that limited railway, waterfront, and local access to the district (Portland Bureau of Planning 2002). Moreover, the waste from industrial activities that had been disposed of into the adjacent lands and water contributed to the South Waterfront's status as a brownfield (Portland Bureau of Planning 2002, p. A-7).

Efforts to revitalise the South Waterfront District began in the late 1970s beginning with the removal of Harbour Drive. The Portland Development Commission (PDC) subsequently acquired 73 acres (30 hectares) within the district in 1978 and made provisions for their redevelopment in the Portland Downtown Urban Renewal Plan (The Center for Brownfields Initiatives 2003). This was followed by the creation of several other long-term renewal plans that addressed the need for urban revitalisation, including the South Waterfront Redevelopment Program (1979), the Central City Plan (1988), the North Macadam Urban Renewal Plan (1999), and most recently the South Waterfront Plan (2002). Since the adoption of the Central City Plan in 1988, the South Waterfront District has been envisioned as an urban mixed-use neighbourhood (Portland Bureau of Planning 2002, p. C-4). While the particular details of the project vision have evolved incrementally to adapt to meet Portland's changing economic and social needs, the primary objective that remained consistent within each successive plan was to re-establish connectivity between the district, neighbouring communities, and the Willamette River.

Visioning for the South Waterfront master plan began in 1997 (US EPA 2012). Several interviewees noted that the details of the master plan vision evolved incrementally through weekly meetings between stakeholders, which fostered consensus and support for the initiative while balancing individual needs and objectives. For instance, Oregon Health and Science University (OHSU) is constrained by its location on Marquam Hill, which rises just west of the South Waterfront site, and was considering opportunities for expansion into Portland's suburbs. The redevelopment of South Waterfront would provide OHSU with land for expansion while acting as a catalyst for economic development in the area. A tram between the existing and planned OHSU campuses was agreed to as part of the vision in order to get faculty, students, staff, and others up and down the hill. The integration of sustainable design, green building, and greening techniques in the master plan vision was also promoted by the stakeholders.

The PDC utilised funding awarded through a $200,000 EPA Brownfields Assessment Project grant to complete a Phase-I area-wide assessment of the property between 2003 and 2004. The EPA facilitated the assessment and remediation efforts in partnership with the Oregon Department of Environmental Quality (DEQ) through its newly established Voluntary Cleanup Program. Several properties did not have any identifiable environmental concerns, whereas others situated mostly in the northern reaches of the district required Phase-II assessments and remediation on a site-by-site basis. The DEQ prescribed a five-year groundwater-monitoring programme in combination with the installation of a river-bank stabilisation system in order to make sure contaminated groundwater near certain projects would not leach into the river.

Phase-I of the South Waterfront redevelopment was initiated in 2003, overlapping with the environmental assessment and remediation efforts, and was initially scheduled for completion in 2008. Phase-I projects are located in the Central District plan area within the South Waterfront District and involve multi-modal transportation infrastructure (to address the physical isolation of the area), greening, housing, and office development (see Figure 2).

Declining water quality and habitat degradation in the Willamette River, as well as a shortage of public space in the South Waterfront District, motivated the city to explore greening efforts that would renew the river ecology while providing waterfront access and park land to the public. Extending the Willamette River Greenway from downtown Portland through the South Waterfront District was intended to facilitate connectivity between the renewal area and surrounding neighbourhoods. Set back an average of 125 feet from the riverfront, the 38-acre (15 hectare) South Waterfront Greenway offers alternative transportation routes as well as passive and active recreational opportunities for visitors

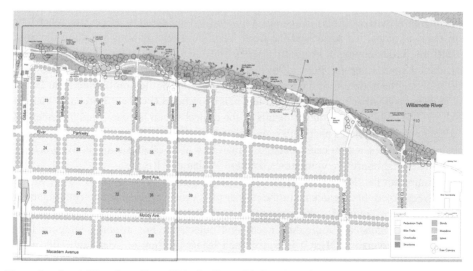

Figure 2. South Waterfront Central District [box added] and portion of the South Waterfront Greenway Site Plan 2004 (courtesy of Portland Parks and Recreation and Walker Macy – Thomas Balsley Associates 2004).

and residents. Development of the greenway has been complicated by the overlapping regulatory permits required to address stormwater runoff and leachate into the river given the industrial heritage of the area. Phase-I of the construction, currently in progress, involves removing contaminated soil and restoring shoreline habitat. The primary source of funding to complete the greenway extension is from public investment at a cost of $10.5 million (Portland Parks and Recreation 2012). A 2-acre (0.8 hectare) neighbourhood park in the Central District, formally known as Elizabeth Caruthers Park, was completed in 2010 on land once utilised as a public storage facility (US EPA 2012). Funding through a $200,000 EPA Brownfields Grant enabled the PDC to assess and remediate the park between 2004 and 2006, with $3.95 million in additional public funding spent to acquire, remediate, and develop the park (US EPA 2012). Remediation involved the removal of existing structures, two underground storage tanks, a railroad spur, and approximately 1250 tons of contaminated soil, with the addition of approximately 7500 cubic yards of clean soil for surface grading (US EPA 2012).

The completed park offers active and passive recreational opportunities while its re-naturalised landscape is designed to absorb and treat stormwater runoff. It is considered a "focal point" for civic activity in the Central District and is part of Portland's open space network.

Urban design guidelines influence the pattern of development within the district, east–west view corridors to the waterfront are preserved and enhanced by requiring buildings to "step back from the street, for the portion of the building over 50 feet in height" (Portland Bureau of Planning 2002, p. C-6). Setbacks for buildings at grade are also required, to a maximum of 12 feet from the sidewalk, in combination with active ground floor uses, including retail, office, and housing that facilitate an attractive and visually interesting pedestrian environment (Portland Bureau of Planning 2002, p. C-6). To achieve a 125-foot setback for the waterfront greenway desired by the city, trade-offs in building height and density had to be made (Portland Bureau of Planning 2002, p. C-6).

The renewal plan for the South Waterfront District includes the development of 3000 residential units that offer market-rate, affordable home ownership, and rental options. Developers are pursuing LEED Neighbourhood Development certification, making the renewal area one of the densest and sustainably designed communities in the USA. Sustainable design strategies include orienting buildings to optimise solar radiation, district heating, mixed-use and transit-oriented development, and on-site stormwater management systems, many of which are very visible bioswales and ecoroofs that aim to filter virtually all of the site's stormwater before it enters the river.

The redevelopment of Portland's South Waterfront District is being financed through a 20-year strategy established by the PDC that leverages public and private resources to generate investment and development in the plan area that would otherwise not be feasible (PDC n.d., p. 7). Given the district's industrial heritage, public intervention was necessary to reduce the environmental and economic costs of development, and to encourage private investment in support of broader policy objectives. A report prepared for the PDC by E.D. Hovee & Company (2003, p. 21) analysing the anticipated return on investment in the South Waterfront District confirms that a public–private approach to urban renewal will allow the project to achieve more of the public policy objectives of the area plan, than would private mechanisms alone. For instance, a market-only approach would have resulted in $352 million of investment compared to the $1.9 billion of public and private investment currently projected by 2020 on build-out. Under this second scenario, $1.7 billion is from private investors with the balance coming from public sources (E.D. Hovee & Company 2003, p. 21). Performance indicators for housing, jobs, transportation, retail, and parks and open space also demonstrate higher quantitative measures through the public–private approach, allowing the redevelopment project to move from a conventional to a more sustainable one.

The public–private partnership is governed by the South Waterfront Central District Project Development Agreement, first signed in 2003 and updated through several subsequent amendments. Public projects are administered and managed by the PDC. Approximately $131 million of the Phase-I public funds are being sourced through Tax Increment Financing (TIF), which prioritises investment in infrastructure such as street and utility construction, extension of the Portland Streetcar, and the riverfront greenway to pave the way for private investment (PDC 2005, p. 1). A significant share of the public funds, $53 million, is allocated for transportation projects, followed by $28 million for affordable housing and parks, and $17 million for additional infrastructure and job investment programmes (E.D. Hovee & Company 2003, p. 10). Private investment, on the other hand, is being directed to the development of office and research facilities, retail establishments, and market-rate housing in the district.

Overall, the desire to develop a large-scale and sustainable mixed-use community allowed for the integration of a broad array of greening elements into the South Waterfront project beyond just a conventional park. These include natural elements like the South Waterfront Greenway, wildlife restoration projects like an Osprey nesting platform, visible green infrastructure through bioswales and ecoroofs, as well as more socially oriented greening via community gardens, a waterfront dog park, and an active nature and green space committee. While the studies discussed above noted that a public–private approach would allow the project to achieve more of the public policy objectives of the area plan than would private mechanisms alone, a few of the interviewees noted that the public greening investments would also have been constrained without the private development interest.

Milwaukee's Menomonee Valley

For almost two decades, greening has played a central role in transforming Milwaukee's Menomonee Valley from one of the most blighted post-industrial districts in the Midwest into a modern green industrial district (De Sousa 2011). The 1400 acre (567 hectare) Valley lies in the heart of the city and flowing through it is the Menomonee River whose abundant marshlands once sustained local Native American populations (Gurda 2003). The Valley's accessibility to water and rail made it a prime location for the industrial activities that started to arrive in the late 1800s after vast quantities of material were used to fill the marsh (Gurda 1999). While the industrial engine of the Valley roared for over a half century, it began to sputter after the Second World War and with its decline came a host of socio-economic and environmental problems.

Interest in revitalising the Valley intensified in the late 1980s. While several long-standing manufacturers continued to operate in the Valley and wanted to see it retain its industrial viability, there also emerged a desire for the renewal of its natural heritage and reconnection to surrounding neighbourhoods. In 1998, the City of Milwaukee put forward a *Market Study, Engineering, and Land Use Plan for the Menomonee Valley* that was a vital step in moving the vision forward. Unlike many plans that aim to clear away an area's industrial legacy, this plan recommended that the area be upgraded and revitalised to retain and strengthen existing industries, attract new industry to the western and central areas of the Valley, promote compatible mixed-use development (largely in the eastern Valley), and maintain and protect adjacent neighbourhoods and business areas. In 1999 the EPA awarded a $250,000 grant to the Sixteenth Street Community Health Center, a local non-profit, through its Sustainable Development Challenge Grant Programme to look into ways of incorporating sustainability into the Valley's redevelopment. The Sixteenth Street Community Health Center (2000, p. 21) organised a two-day charrette in which design professionals, nonprofits, government agencies, local universities, students, and community members were charged with the task of "raising the bar on redevelopment and restoration activities for Milwaukee's Menomonee River Valley". The charrette resulted in the production of a comprehensive report entitled *Vision for Smart Growth* that outlines ideas for the eastern, central, and western portions of the Valley. The plans and designs provided a portrayal of the Valley's assets and potential in terms of greening and sustainability that became embraced by stakeholders.

Although the planning, economic development, and infrastructure stories of the Valley's redevelopment are insightful, the central role that greening has played in the resurrection of this post-industrial district is of considerable interest because it is arguably the main feature that has changed the perception of the area and has allowed it to be woven back into the fabric of the city. Three key projects include the Hank Aaron State Trail, The Menomonee Valley Community Park, and the Former Airline Yards Property (Figure 3).

Despite its central location, one of the key criticisms of the Valley was that it was extremely difficult to get into. Several major projects were initiated in the 1990s to improve access and mobility within the Menomonee Valley, including the Hank Aaron State Trail, Miller Park, the Sixth Street Viaduct, and Canal Street. While the individual projects are each important, more significant was the fact that the group of infrastructure projects brought the collective attention and resources of numerous stakeholders and government agencies to the area. In 1991, the Wisconsin State Legislature directed the Department of Natural Resources (DNR) to examine the feasibility of establishing a Hank Aaron State Park on the Menomonee River adjacent to Milwaukee County Stadium (Wisconsin DNR 1996). More comprehensive planning for a greenway trail to connect the Valley from

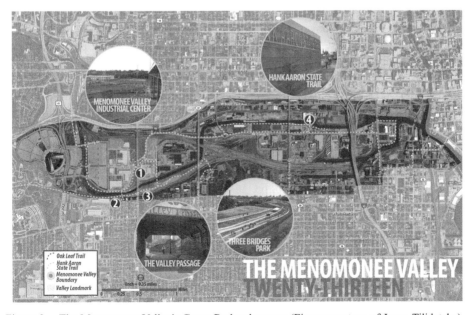

Figure 3.   The Menomonee Valley's Green Redevelopment (Figure courtesy of Jason Tilidetzke).

west to east was initiated in 1992 with the DNR taking the lead in planning, constructing, and managing the trail. Other partners included the City of Milwaukee (involved primarily in raising funds, releasing land, and maintaining the trail), various federal agencies, local community groups and neighbourhood associations (e.g. Friends of the Hank Aaron State Trail group), and private landowners (e.g. Miller Park Stadium Corporation and The Sigma Group donated easements for the trail and helped finance development and re-naturalisation activities). The state trail, Wisconsin's first in an urban area, officially opened in 2000 on the Valley's west side and was connected to Sixth Street in the Valley's east end by 2007. The greening of the western portion of the trail was interwoven with the development of Miller Park Stadium, home of the Milwaukee Brewers, thus giving over two million annual baseball fans a glimpse of the Valley's ongoing green transform-ation. In 2012, another 2.5 miles was added to the Hank Aaron State Trail to connect it with the extensive Oak Leaf Trail that encircles Milwaukee County. The Friends of Hank Aaron State Trail continues to be the primary advocate of the Trail, working closely with the Wisconsin DNR and local partners to plan future portions of the trail and to organise programmes (e.g. river cleanups, plantings, a 5 k Run/Walk, and so-called hikes through history).

Redevelopment of the 140-acre (57 hectare) Milwaukee Railroad Shops property located in the western end of the Valley into a modern industrial centre provided an oppor-tunity for stakeholders to convert sustainable visions, designs, and guidelines into reality. Sixteenth Street Community Health Center (2002), city government, and other sponsors organised a national *Natural Landscapes for Living Communities* design competition in 2001 to plan the redevelopment and greening of the site even before it had been acquired by the city through eminent domain action. The land use, infrastructure, and sustainability visions that evolved from the charrette became entrenched as criteria presented to design teams competing to develop a final site plan. The winning design put forward in 2002 pro-vided for 70 acres (28 hectares) of light industrial development, a mile segment of the Hank Aaron State Trail, and 70 acres of streets, parks, and natural areas along the banks of the river. From this design, the City of Milwaukee generated the *Menomonee Valley Industrial Center and Community Park Land Use Plan* in 2006 to guide redevelop-ment. To make the site "shovel ready" for redevelopment, the City established a $16 million dollar TIF District in 2004 to pay the cost of site remediation, demolition, filling and grading, stormwater utilities, local roadways, and infrastructure. The site required massive cleanup, demolition, removal, and management of six miles of brick sewers, asbestos, and over a million square feet (93,000 m^2) of old building foundations, as well as the trucking of 700,000 cubic yards of fill from a nearby interchange project to create an environmental cap that would protect human health and the environment and raise the site out of the flood plain. The City of Milwaukee aggressively raised funds for remediation and redevelopment activities, including the park development, winning more than 20 local, state, and federal grants and dozens of private donations totalling $24 million (Misky and Nemke 2010, pp. 14–15).

The *Community Park* portion of this project provides an amenity for businesses located in the Valley as well as green space for local residents. Material reuse has been an important component of the park's development linking it to its industrial past, including building debris used to create landscaped mounds, highway project fill to raise the site, crushed con-crete to build stormwater conveyance structures, glass to make pathway railings, beams to build park benches and tables, and so forth (Misky and Nemke 2010). The stormwater portion of the park provides attractive and functional infrastructure for conveying, storing, and treating stormwater for the adjacent parcels in the new industrial area, as well as for Canal Street and other internal roads. The shared stormwater facility makes it unnecessary for developers with parcels in the industrial area to set aside land and build their own small and typically unusable private detention ponds, saving them money and also allowing the city to maximise the build-out of the site. Runoff from the site is routed through a system of subsurface pipes, settles in shallow wetland meadows, and is then filtered through a layer of crushed concrete. Gravity then takes the runoff further down the line to the Swamp Forest section of the park, where natural vegetation absorbs many remaining contaminants. The Wisconsin DNR (2009) notes that the entire process removes more than 80% of total suspended solids as well as phosphorous, nitrogen, and heavy metals. Annual management costs are shared through fees among individual business owners in the industrial centre and the City of Milwaukee. Other parts of the park contain ball fields, public art, portions of the Hank Aaron Trail, and a canoe launch.

As the industrial centre took shape, stakeholders started to work together across the river to the south on yet another project aimed at transforming a vacant railroad switching yard, known as Airline Yards, into a 24-acre (9.7 hectare) public park. In 2011, a development plan was completed for the site that included a mile of trails, water-quality improvements, native landscape restoration, community gardens, and two new bridges to connect the site to residents south of the Valley who have had limited access to it for decades (Menomonee Valley Partners 2012). The park is part of the Hank Aaron State Trail and serves as an outdoor classroom for the Urban Ecology Center, a non-profit environmental education and stewardship group that recharges urban green space through daily education and community programmes. The Wisconsin Bicycle Federation also moved its Milwaukee offices into the Pedal Milwaukee Building, just off the recently constructed Valley Passage Bridge on the Hank Aaron State Trail. The trail construction is being managed by the Wisconsin Department of Transportation and is slated for completion in 2013. The $10.6 million project is being coordinated by Menomonee Valley Partners and Urban Ecology Center in partnership with the State of Wisconsin and the City of Milwaukee (Schlabowske 2012, p. 4). The funds for the project have come from a mix of public-sector grants as well as foundations, corporations, and individuals.

Benefits, barriers, and lessons

Achieving the multiple benefits associated with greening is an increasingly important objective for many cities, especially densely populated ones where opportunities can only be found in a sea of brown and grey spaces. In interviews conducted by the author for the present case studies and in past research on other brownfield to park conversion projects, both implementers and users enumerate a range of benefits from these projects. Implementers interviewed in the USA highlight the provision of new open spaces and recreational sites, economic benefits (especially tax revenue generation onsite and via raising the value and marketability of surrounding property), aesthetic benefits (especially blight elimination), and the conservation and the restoration of habitat (De Sousa 2004). Other benefits often identified include the interconnection of the newly formed green spaces to other parts of the city, the creation of new trails, new access venues to water, flood control, infrastructure improvement, cohesion among the neighbourhood residents, historical restoration, and providing models for future greening efforts. Park users surveyed by the author note *personal* benefits, such as physical activity, children's play, scenic beauty, relaxation, access to nature, and social interaction, as well as *community benefits* that focus on scenic beauty, trails for walking, neighbourhood appeal, access to nature and recreation space, community pride, blight removal, and public fitness (De Sousa 2006). An interesting finding from surveys of park users was that many did not even remember what had been there before the green space, pointing out that they just did not go to the area until people told them about the green transformation that had occurred. All in all, it is clear that the objectives and benefits associated with greening projects are broad-based and reflective of the Sustainable Park model put forward by Cranz and Boland (2004), regardless of whether the greening is achieved via a park development or mixed into a larger sustainable development as witnessed in the case studies above.

Despite these benefits, the greening of post-industrial districts is hindered by a variety of real and perceived obstacles. The single most challenging barrier continues to be a lack of financial resources for planning, coordinating, and undertaking remediation and green redevelopment. Other factors impeding the conversion of brownfields-to-parks specifically often include land acquisition (particularly along linear spaces), concerns about soil contamination, the lack of government leadership, staff expertise, stakeholder trust, alternative redevelopment priorities, and concerns about long-term maintenance costs. It is for this reason that incorporating greening into sustainable city and sustainable development projects more broadly and both involving and leveraging the resources of other stakeholders may be a more feasible approach in the long run. While current research by the author examining sustainable brownfield best practice projects is revealing that many barriers to greening projects also surface in sustainable redevelopments (namely financial costs, technical capacity to incorporate green strategies, local regulatory/permitting restrictions, negotiation, and educating stakeholders on the utility of a more sustainable approach), the ability to infuse various types of greening into the planning conversation and leverage the attention and resources of a more comprehensive effort make it a more viable option than keeping parks as an isolated land use. This leads to a number of lessons that will hopefully assist those willing to take on the greening challenge:

- Greening is particularly effective when all parties can link green space needs with brownfield site availability as part of a comprehensive urban revitalisation plan that considers land use, infrastructure, and sustainability;
- The involvement of communities in the redevelopment process is crucial, in both the short and long term, particularly for stand-alone brownfield-to-park projects;
- Early planning efforts should be funded and carried out to devise and integrate sustainability and greening visions and plans in a manner that involves multi-stakeholder participation and innovative participatory techniques;
- The greening of post-industrial districts seems to be most feasible when it is justified on the basis of multiple objectives that can include recreation, habitat, green infrastructure, education, stewardship, enhancing visual appeal, and reconnecting the site to the community;
- A phased approach to remediation and development that also integrates elements of landscape design with site remediation should be used to enhance the feasibility of greening projects;
- Promoting greening as part of a broader public–private partnership project is more viable than a stand-alone public park project because it brings together the collective attention and resources of numerous stakeholders; and
- Park departments and others interested in greening should devote more attention to understanding how greening can be used to leverage private investment and development because the opportunity for private investment allows for an increase in and legitimises the public expense for greening.

Acknowledgements

The case studies research was performed under a subcontract with the University of Illinois at Chicago and made possible by grant number TR-83418401 from US EPA and its contents are solely the responsibility of the author and do not necessarily represent the official views of the University of Illinois. The author would like to sincerely thank the interviewees from Milwaukee, Portland, and New York for graciously providing a wealth of information about the projects. The author would also like to thank his student assistants – Lily-Ann D'Souza, Kevin Duffy, Jason Tilidetzke, Marta Brocki, Laura Lynn Roedl, Tim Streitz, and Elizabeth Durkin – for their research support.

References

Anon., 2009. Gas Tank Park: Queens creates green space at former gas storage facility. *New York Construction* [online], May. New York: McGraw Hill Construction. Available from: http://newyork.construction.com/features/archive/2009/05_f2b_elmhurstpark.asp [Accessed 23 September 2012].

Bradshaw, A., 2000. The use of natural processes in reclamation – advantages and difficulties. *Landscape and Urban Planning*, 51 (2–4), 89–100.

Bunster-Ossa, I.F., 2001. Landscape urbanism. *Urban Land*, July, 37–47.

Burger, J., *et al.*, 2004. Assessing ecological resources for remediation and future land uses on contaminated lands. *Environmental Management*, 34 (1), 1–10.

Byrne, J., Kendrick, M., and Sroaf, D., 2007. The park made of oil: towards a historical political ecology of the Kenneth Hahn State Recreation Area. *Local Environment: The International Journal of Justice and Sustainability*, 12 (2), 153–181.

The Center for Brownfields Initiatives, 2003. *EPA region 10 – South Waterfront redevelopment project, Portland, Oregon.* New Orleans: University of New Orleans.City of Milwaukee, 1998. *Market study, engineering, and land use plan for the Menomonee Valley.* Report prepared for the Department of City Development, City of Milwaukee. Milwaukee, WI: Lockwood Greene Consulting, Fluor Daniel Consulting, Trkla, Pettigrew, Allen, and Payne, Inc., and Edwards and Associates.

City of Milwaukee, 2006. *Menomonee Valley Industrial Center and Community Park Master Land Use Plan.* Report prepared by the City of Milwaukee, Redevelopment Authority of the City of Milwaukee. Milwaukee, WI: Department of City Development.

City of New York, 2011. *plaNYC: A Greener, Greater New York.* Update April 2011. New York: City of New York [Brownfields Section, Initiative 11 – Promote green space on remediated brownfield properties].

City of New York Parks and Recreation, 2010. *High performance guidelines: 21st century parks for NYC.* New York: Design Trust for Public Space and the City of New York acting by and through the New York City Department of Parks and Recreation.

CL:AIRE (Contaminated Land: Applications in Real Environments), 2009. *Integrated remediation, reclamation and greenspace creation on brownfield land.* Surry: Centre for Forestry and Climate Change, subr:im bulletin 11 (April 2009). Available from: http://www.forestry.gov.uk/pdf/SUBRIM_bulletin_11.pdf/$file/SUBRIM_bulletin_11.pdf [Accessed 16 August 2013].

Cranz, G., 1982. *Politics of park design.* Cambridge, MA: The MIT Press.

Cranz, G., and Boland, M., 2004. Defining the sustainable park: a fifth model for urban parks. *Landscape Journal*, 23 (2), 102–120.

De Sousa, C., 2003. Turning brownfields into green space in the City of Toronto. *Landscape and Urban Planning*, 62 (4), 181–198.

De Sousa, C., 2004. The greening of brownfields in American cities. *Journal of Environmental Planning and Management*, 47 (4), 579–600.

De Sousa, C., 2006. Unearthing the benefits of brownfield to green space projects: an examination of project use and quality of life impacts. *Local Environment: The International Journal of Justice and Sustainability*, 11 (5), 577–600.

De Sousa, C., 2010. From brown liability to green opportunity: reinventing urban landscapes. *Carolina Planning*, 35, 3–12.

De Sousa, C., 2011. Creating the green industrial district: transforming Milwaukee's Menomonee Valley from a blighted brownfield into a sustainable place to work and play. In: M. Slavin, ed.

Sustainability in America's cities: creating the green metropolis. Washington, DC: Island Press, 45–68.

E.D. Hovee & Company, 2003. *North Macadam urban renewal area return on investment (ROI) analysis update*. Report prepared for the Portland Development Commission, Vancouver, Washington.

Eisen, J.B., 1999. Brownfields policies for sustainable cities. *Duke Environmental Law and Policy Forum*, 9 (2), 187–229.

Greene, C., 2008. Elmhurst gas tanks park taking shape. *The Forum* [online], October 2. Available from: http://forumnewsgroup.blogspot.ca/2008/10/elmhurst-gas-tanks-park-taking-shape.html [Accessed 9 July 2012].

Gurda, J., 1999. *The making of Milwaukee*. Milwaukee, WI: Milwaukee County Historical Society.

Gurda, J., 2003. *The Menomonee Valley: a historical overview*. Milwaukee, WI: Report prepared for the Menomonee Valley Partners.

Harnik, P., 2000. *Inside city parks*. Washington, DC: Urban Land Institute and the Trust for Public Land.

Hofmann, M., Westermann, J., and Kowarik, I., 2012. Perceptions of parks and urban derelict land by landscape planners and residents. *Urban Forestry and Urban Greening*, 11 (3): 303–312.

Hughes, C.J., 2006. Polyglot haven awaits a 'Gas Tank Park'. *The New York Times* [online], 3 September. Available from: http://www.nytimes.com/2006/09/03/realestate/03livi.html?page wanted=all [Accessed 15 July 2012].

ICMA (International City/County Management Association), 2002. *Growing greener: revitalizing brownfields into green space*. Washington, DC: International City/County Management Association.

Juniper Park Civic Association, 2003. Timeline of the Elmhurst gas tanks property. *Juniper Berry* [online], December. Available from: http://www.junipercivic.com/juniperberryarticle.asp? nid=202 [Accessed 23 July 2012].

Lopes, P., 2011. *Mayor Bloomberg opens Elmhurst Park in Queens* [online]. New York City.gov. Available from: http://www.nyc.gov/cgi-bin/misc/pfprinter.cgi?action=print&sitename=OM &p=1348756030000 [Accessed 23 July 2012].

Loures, L., and Panagopoulos, T., 2007. Sustainable reclamation of industrial areas in urban land-scapes. *WIT Transactions on Ecology and the Environment, Sustainable Development and Planning III*, 102: 791–800.

Menomonee Valley Partners, 2012. *Making a collective impact, 2011 annual report*. Milwaukee, WI: Menomonee Valley Partners.

Misky, D., and Nemke, C., 2010. From blighted to beautiful. Government Engineering, May–June, 14–16.

New York State Department of Environmental Conservation, 2012. *Newtown – Elmhurst former gas holder* [online]. New York: New York City Department of Environmental Conservation. Available from: http://www.dec.ny.gov/chemical/53590.html [Accessed 23 July 2012].

PDC (Portland Development Commission), 2005. *South Waterfront Central District diversity in con-tracting and workforce training report fiscal year 03/04*. Portland: Portland Development Commission.

PDC (Portland Development Commission), n.d. *South Waterfront District FAQS*. Portland: Portland Development Commission.

Platt, L., 2010. Planning ideology and geographic thought in the early twentieth century: Charles Whitnall's progressive era park designs for socialist Milwaukee. *Journal of Urban History*, 36 (6): 771–791.

Portland Bureau of Planning, 2002. *South waterfront plan*. Portland: City of Portland.

Portland Parks and Recreation, 2012. *South Waterfront Greenway*. Portland: City of Portland. Accessed from: http://www.portlandonline.com/parks/index.cfm?c=45643 [Accessed 9 July 2012].

Portland Parks and Recreation and Walker Macy – Thomas Balsley Associates, 2004. South Waterfront Greenway Site Plan 2004 [online]. [Accessed with permission on 19 November 2013].

Schilling, J., and Logan, J., 2008. Greening the rust belt: a green infrastructure model for right sizing America's shrinking cities. *Journal of the American Planning Association*, 74 (4), 451–466.

Schlabowske, D., 2012. Airline Yards another home run for Hank Aaron. *The Bicycle Blog of Wisconsin* [online], August 30, 1–6. Available from: http://bfw.org/2012/08/30/airline-yards-another-home-run-for-hank-aaron/ [Accessed 16 August 2013].

Siikamäki, J., and Wernstedt, K., 2008. Turning brownfields into greenspaces: examining incentives and barriers to revitalization. *Journal of Health Politics, Policy and Law*, 33 (3), 559–593.

Sixteenth Street Community Health Center, 2000. *A vision for smart growth: sustainable development design charrette Milwaukee's Menomonee River Valley 1999–2000*. Milwaukee, WI: Sixteenth Street Community Health Center, Department of Environmental Health.

Sixteenth Street Community Health Center, 2002. *Menomonee river valley national design competition, executive summary*. Milwaukee, WI: Competition sponsored by the Sixteenth Street Community Health Center, Menomonee Valley Partners Inc., the City of Milwaukee, the Milwaukee Metropolitan Sewerage District, Wisconsin Department of Natural Resources and Milwaukee County.

Trust for Public Land, 2013. *ParkScore index* [online]. Washington, DC: Trust for Public Land. Available from: http://parkscore.tpl.org/ [Accessed 16 August 2013].

US EPA, 2012. *Brownfields at-a-glance – Elizabeth Caruthers Park* [online]. Seattle, WA: EPA Region 10. Available from: http://www.epa.gov/brownfields/success/portland_or_caruthers_brag.pdf [Accessed 9 July 2012].

Wisconsin DNR, 1996. *Henry Aaron State trail, feasibility study master plan and environmental assessment*. Madison, WI: Menomonee Valley Greenway Advisory Committee, National Park Service, and Wisconsin Department of Natural Resources.

Wisconsin DNR, 2009. *Menomonee valley –stormwater park*. Madison, WI: Wisconsin Department of Natural Resources Remediation and Redevelopment Program, PUB-RR-827.

Environmental gentrification in a post-industrial landscape: the case of the Limhamn quarry, Malmö, Sweden

L. Anders Sandberg

Faculty of Environmental Studies, York University, HNES Building, Toronto, ON, Canada

The paper uses the lens of environmental gentrification and post-industrial landscapes to explore a limestone quarry converted into a nature reserve, the Limhamn quarry, Kalkbrottet, in Malmö, Sweden. The research is based on field investigations, a review of the primary material, and interviews of key actors. The findings suggest that the quarry has been subjected to processes of environmental gentrification threatening to make the quarry into a gated ecology. City ecologists, drawing on national traditions in support of common green space, working with a spontaneously appearing unique flora and fauna, have countered the environmental gentrification process by seeking nature protection status of and public access to the quarry. The paper suggests that by more fully integrating the history of industrial work, rogue subjects who now frequent the quarry illegally, and new immigrants who may find a familiar physical landscape in the quarry, the site could become meeting place for "others" and force against environmental gentrification.

Typically, post-industrial landscapes are associated with places in disuse, abandonment, and disrepair, such as old factories, harbours, train yards, tenements and barracks, mines, and sites at the edges of transportations routes, in short, "places that look empty, and appear as ones which do not have any use (any more)" (Doron 2000, p. 247). They may also be connected with poor or marginalised communities that were once compelled to accept to live in industrially polluted or polluting areas because of cheaper rents and real estate or because they could not resist the siting of such industries in their neighbourhoods. The literature on post-industrial landscapes, however, proposes that such landscapes are not doomed to perpetual contamination and devastation (Kirkwood 2001). Nature can heal itself in surprising ways and ecological restoration professionals can both help the process along as well as create new landscapes with novel ecologies. When restored, however, post-industrial landscapes may be subject to gentrification, mainly because of their prime location near downtowns, waterfronts, or rivers. One form of such gentrification, environmental gentrification, constitutes a situation where old factory sites, quarries, landfills, and railway corridors, sometimes referred to as brownfields, are transformed into green areas with high amenity values. The connection between the greening of post-industrial sites and gentrification can be fraught with contestation and the unique aspects of place and time. In some cases, the greening process can benefit existing poor communities (Groth and Corijn 2005, Lund Hansen 2010), in other situations it may not (Foster 2010, Checker 2011, Bryson 2012), and in yet other places, the impact may be contradictory (Curran and Hamilton 2012). In this paper, I interrogate the environmental gentrification in one post-industrial landscape, a limestone quarry in Limhamn, Malmö, hereafter Kalkbrottet, where industrial-scale extraction started in 1863 and ceased in 1994. Sixteen years later, in 2010, the quarry was a full-fledged nature reserve sanctioned by municipal, national, and international authorities. It was then over 90 ha, 4 km in circumference, 70 m deep at its lowest point, and surrounded by gentrified and gentrifying communities (Figure 1).

I base my account on various local industrial and local histories of Malmö, city official plans, media stories, several interviews of some of the principal actors in the emergence of the quarry as a nature reserve, and personal observations visiting the quarry, and walking its surroundings. Nine subjects were interviewed in June 2012, including the city biologist who spearheaded the effort to designate the quarry a nature reserve, two local Green Party politicians, an environmentalist in Limhamn, a member of a local industrial heritage organisation, a long-term retired senior employee for the quarry company, and three academics (two historians and an anthropologist), with an intimate knowledge of the history of Malmö and the social history of Limhamn.

Particular analytical themes of concern relevant to environmental gentrification include the policy and planning framework, representational practices associated with real estate marketing, everyday practices of local residents, public access, working class heritage,

Figure 1.   Kalkbrottet looking west towards the bridge that connects Sweden and Denmark.
Source: Photo courtesy of the author.

social disparities, immigration, and the role of the non-human in people-environment dynamics. I am particularly interested in exploring the unique role of place, history, and nature under the pressure of neoliberalism, in this case Sweden and the City of Malmö, as (f)actors at work in influencing the greening and gentrification processes, pressures and challenges at Kalkbrottet.

Post-industrial landscapes and environmental gentrification

There is now a substantial literature on post-industrial landscapes and their transformations (Kirkwood 2001). Rather than being seen as abject or sacrificial spaces laden with pollutants and devoid of resources, they are typically appreciated for the capital accumulation potential associated with intensive redevelopment. In some cases, however, it is their social but especially ecological potential that drives interest. Germany's Duisburg North Park, New York City's Highline, Staten Island's Fresh Kills, and the Denver Arsenal constitute some famous cases (Herrington 2009, pp. 77–82).

The ecological restoration literature on post-industrial landscapes seldom addresses environmental gentrification issues, and, when it does, it is seldom the main focus. Eric Higgs, for example, provides a series of best practice criteria that include ecological integrity and wild design, and then includes a couple of measures that can address environmental justice issues: historical fidelity (paying attention to previous uses) and focal practice (including all stakeholders in the planning and execution of the project; Higgs 2003). Such criteria are easily, however, co-opted by powerful property and expert actors who pose as local interests (Higgs 2003). A growing literature describes areas being rebuilt, repurposed, and greened and pushing older typically low-income, homeless, and perhaps racialised minorities out while inviting well-paid professionals of the "creative class" and well-to-do retirees to take their place, resulting in "gated ecologies" (Dooling 2009, Cunningham 2012, p. 153). Some of the post-industrial landscapes identified in the preceding paragraph, such as High Line and Duisburg North Park bear the distinct traits of environmental gentrification (Storm 2008). There are recent studies in Toronto, New York City, and Spokane that show the ability of gentrifiers to use greening strategies to serve their own interests (Foster 2005, 2010, Checker 2011, Bryson 2012).

But there are also studies that celebrate resistances to environmental gentrification in post-industrial landscapes. Groth and Corijn (2005) refer to informal actors reclaiming urbanity, with varying degrees of success, in "indeterminate spaces" like the Makiasiinit warehouse complex in the central parts of Helsinki, the Raw-Tempel railway shops in Berlin, and the Léopold Railway Station in Brussels. Copenhagen's Christiania, Europe's longest existing squatter settlement, constitutes another example of social and ecological transformation (Lund Hansen 2010). At Greenpoint, Brooklyn, local groups have employed the strategy of "just green enough" to counter the pressures of environmental gentrification, an initiative helped by the mortgage crisis in the USA housing market (Curran and Hamilton 2012).

Conceptualising these environments as contingent ecologies, where human and non-human processes act out or are permitted to act out in landscape change, helps open up a reflexive space for resisting the process of environmental gentrification. This typically involves local resident groups. But it may also encompass state institutions, expert and political actors, and national customs which can temper the impact of environmental gentrification, a situation that contrasts with many other jurisdictions where the very same actors have subverted grassroots greening initiatives in favour of gentrification (Newman 2011). In Sweden, such institutions, experts, and customs include the national tradition of bureaucratic and expert rule (Lindvall and Rothstein 2006); the open access to non-human nature in the shape of *allemansrätten,* which permits Swedes and visitors to access non-human natures, even privately owned lands, as long as the privacy of the owner is not disturbed. The custom flows from a number of historical and geographical conditions and is more permissive than in most other countries; and the long-standing state policy promoting (recreational) life in the open air for the masses, *friluftsliv* (Gelter 2000). More-over, following Latour's actor-network theory, the role of non-human actors may influence the environmental gentrification process. According to Latour (2005), an actor can be any-thing non-human that alters a situation, therefore changing others' actions. Ecological and geological processes, in this situation, constitute a politics by other means. As Qviström and Saltzman (2006, p. 22) put it, ecologies and their associated landscapes, flora, fauna, and scenes can "offer themselves up to different and competing interpretations and interests, and every activity and representation will either question or confirm the dominant ways of seeing and ways of acting within the landscape". Post-industrial sites are often tied up with floral and faunal urban ecologies that correspond to their social ecologies. There is clearly a mutual dynamic between ecological and social transformations, where, for example, certain flora and fauna are associated with the gentrification process, while other flora and fauna are connected to other types of social formations. Dense and lush vegetation, for example, may be amenable to sleeping homeless people and gay men's sexual activities while an open and orderly vegetation pattern may be associated with the gentrification process (Doron 2000, Hobbs and Sandilands 2013). Often the natures that the gentrification process displaces are seen as random and disordered wastelands, but, as Doron states, the nature and activities that take place in such wastelands are "simply an order of a different kind" (Doron 2000, p. 249).

Malmö: a gentrifying city

The extraction of limestone in Limhamn area started in the early seventeenth century, but industrial-scale operations did not begin until the second half of the nineteenth century. Extraction started at Kalkbrottet in 1863 and it was at one point the largest limestone quarry in Sweden. In 1890, the operations were complemented by cement manufacturing in close vicinity. Both operations were owned locally but were in time absorbed by a nation-ally based company and bought in 1998 by the German-based multinational corporation HeidelbergCement. In the process, limestone extraction and cement production were con-fined to bigger operations in fewer locations; consequently, the Limhamn cement plant was closed in 1978, while Kalkbrottet ceased operations in 1994.

Limhamn started as a fishing village and peasant small-holding community but with the expansion of the quarry, and the addition of the cement factory and ancillary industries, the village evolved rapidly into a working class town. Capital-labour relations were paternalis-tic. The industrial patron ruled his workers with a firm hand, while providing them with housing, garden plots, and funds for religious activities (Lundin 2006). Then, in 1915, when the benevolent entrepreneur died and the town had grown to 10,000, Limhamn became part of the City of Malmö.

A close relationship emerged between the cement capital in Limhamn, and the industrial and business elite, the social democratic state and the labour movement in Malmö. Journal-ist Svensson (2010) refers to Kalkbrottet as the uterus of modern Malmö, an industry that helped Malmö retain its position as Sweden's third largest city, a rank it has held since 1870. For well over a century, Malmö remained an industrial city revolving around shipbuilding, secondary manufacturing products, cement, and construction material, and ruled by strong social democratic traditions. The Swedish transnational development and construction company Skanska remains a legacy of Kalkbrottet and the cement production in Limhamn. Malmö was the classic social democratic city, where the capital-labour compromise resulted in relative peace on the labour market, high wages, full employment, and welfare provisions that included a public sector that ensured accessibility to high-quality housing at modest costs to the working class.

In the 1980s, however, the industrial economy of Malmö was hit hard. The Kockum shipyards, the largest employer, closed operations and Malmö's population declined. To reverse its fortunes, the city government adopted a new strategy. Baeten (2012) labels the new planning scenario a mongrel or hybrid neoliberalism. On the one hand, the city government adopted a development policy with the typical traits of neoliberalism, the emphasis on market mechanisms in promoting growth, the spectacularisation of architectural and infrastructural projects, and the gentrification of former middle and working class neighbourhoods to accommodate the creative class (Jamison 2008, Koglin 2008, Baeten 2011, 2012). Spectacularisation is an extreme form of commodification, marketisation, and branding. It has to do with visualisation, the creation, and recognition of the spectacular, making it into a commodity that can be sold itself (as well as a means to commodify its surroundings), and that is divorced from or distorts the processes and histories that made it (Lefebvre 1996, Debord 2006). In the current city, the spectacular typically includes buildings, sports stadiums, museums, and mega-events, but it can also include post-industrial landscapes in the form of naturalised industrial areas that promote biodiversity and other green objectives.

The so-called Örespectacle is part of an effort to connect Malmö with the capital city of Denmark, Copenhagen, creating a bi-national city region, Örestad, named after Öresund, the strait that divides Sweden and Denmark (Figure 2). The strait is since 1999 connected by a bridge. This project, Baeten argues, is a function of an interventionist state providing the necessary financial and technical support to complete the project. Baeten (2012) likens the process to a form of Keynesianism for the wealthy rather than for the poor and middle class.

Harvey (2005, p. 115) has labelled Sweden an example of "circumscribed neoliberalism", a neoliberalism with a superior social condition in comparison to other countries. Clark and Johnson (2009), however, have shown that in some sectors, such as housing, social housing has more or less been replaced by privatisation schemes (Hedin *et al.* 2012). In Malmö, the former waterfront area that harboured industries (including Limhamn), is now subject to a process of gentrification. The most prominent example is Västra Hamnen, the Western Harbour, composed of expensive condominiums and high-cost rental housing units, an area that previously constituted the shipbuilding area of the City, and whose skyline was dominated by the huge cranes of a busy port (Figure 2). The area is anchored by Turning Torso, the twisted residential tower that was drawn by a Spanish (st)architect Santiago Calatrava and sponsored by HSB, the Swedish Tenant-Owner Cooperative Housing Association. The Turning Torso constitutes an attempt to use a spectacular landmark to put Malmö on the world map (for similar attempts in other contexts, see Belanger 2000, Lehrer 2003, Broudehoux 2010, Igoe 2011). At Hyllievång, a community a mere kilometre to the east of Kalkbrottet, Baeten (n.d.) describes a similar situation of top-down planning by an exclusive group of politicians, senior planners, and developers where a wealthy population, including people who work in Denmark, is targeted through the provision of top end retail and entertainment facilities (Figure 2). The development represents a "postmodern" exclusive architectural scheme reminiscent of Turning Torso at Västra Hamnen (Baeten n.d.).

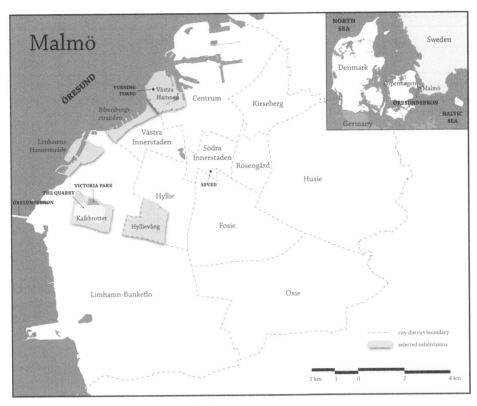

Figure 2.   Map of Malmö. Map by Rajiv Rawat.

Meanwhile, homelessness is growing rapidly and low-income families find it increasingly difficult to locate housing. In addition, the large number of immigrants that Malmö has welcomed in the last two decades is mainly confined to a poor housing stock in the eastern parts of the city. Visible minorities now constitute over 20% of the population, many of them Muslims from southern Europe and the Middle East (Salonen 2012). This has resulted in a more class- and ethnically divided city, and tensions between cultural groups. The disparity between the "two" cities is seldom problematised as a collective and inter-connected problem. Instead, the recent immigrant communities, such as Rosengård and Seved, are singled out as isolated and separate problem areas (Figure 2). By contrast, the lack of diversity in the wealthy parts of Malmö remains unexamined (Lundén 2010).

Kalkbrottet: a case of environmental gentrification?

The local state has been instrumental in rolling out services to make the land available for the gentrification process. The area surrounding Kalkbrottet sits in the privileged regions of Malmö that have undergone and is undergoing a process of gentrification. Limhamn is no longer a working class city but a wealthy part of Malmö, a part of the city inhabited by well-established white ethnic Swedes who enjoy a highly homogenous social environment.

Table 1 shows an example of the deep and increasing divisions within Malmö by comparing Limhamn-Bunkeflo with the city of Malmö as a whole and an immigrant district,

Table 1.   Indicators of gentrification in Malmö.

	Malmö		Limhamn-Bunkeflo		Rosengård	
	1996	2008	1996	2008	1996	2008
Disposable average income (Swedish Crowns)	146,000 (1993)	226,140 (2006)	201,200 (1993)	357,705 (2006)	125,400 (1993)	180,061 (2006)
Percentage on social assistance	15	8 (2007)	2	1 (2007)	48	27 (2007)
Rental properties	51	47	23	25	69	68
Foreign background	27	28	11	14	74	60
Number of cars per 100 residents	29 (1995)	29 (2006)	39 (1995)	37 (2006)	17 (1995)	19 (2006)
Post-high school education	22 (1993)	40	34	48	7 (1993)	18

Source: Malmö Stad, various dates.

Rosengård. Kalkbrottet is located in Limhamn-Bunkeflo, a part of the city with the highest incomes, highest income increases, lowest number of people on social assistance, lowest percentage of rental properties (mainly owned by private companies rather than the local state), and lowest percentage of people of foreign background (who are mainly Danes and other Europeans with professional jobs).

Kalkbrottet and the cement factory grounds (Limhamns Hamnområde) still constitute areas of "no data" in the census tracts because residences have only recently been built or are in the process of being built (Figure 2). But other data indicate that both areas are entangled in an environmental gentrification at several scales. At the level of the city itself, Kalkbrottet figures prominently in tourism advertisements, and is marketed in public discourse and sold as a spectacular place to visit. At the City's website, the quarry is described as "... one of the Öresund region's mightiest places and people commonly call it Malmö's Grand Canyon" (Malmö Stad 2012). Such depictions emphasise the scenic rather than unique ecological characteristics of Kalkbrottet, likely because public access is confined to viewing the quarry from the edge and sporadic supervised tours into the quarry itself. Kalkbrottet and Turning Torso are spectacles that promote a new Malmö (Figure 3). As Journalist Svensson (2010) evocatively states: "From the bottom of a 60 meter deep hole, Malmö has climbed to the top of the roof of a 190 meter high tower. There she stands and she's the best".

After the closure, Kalkbrottet remained in limbo. But by 2006, the owner, Heidelberg-Cement, launched an initiative to exploit the building potential of the surrounding area (Olsson 2007). The City of Malmö took over Kalkbrottet and its maintenance (which included the high cost of continuously pumping out water to keep the quarry from flooding), while the company was granted the right to sell building permits for about 160,000 m^2 of residential floor space (Ivarsson 2007). The interactions and transactions resemble Baeten's characterisation of the City of Malmö as a facilitator and booster of private housing schemes in the city.

The housing surrounding Kalkbrottet is now a combination of rental and condominium units and single-family-homes that target upper-middle and high-income groups. The developments surrounding the quarry are based on the principle of Le Corbusier's concept of a "house in the park" but without the high density and catering to automobile use. Such an arrangement, according to one source, wastes space and resources, and inhibits

Figure 3.   Kalkbrottet looking north towards Turning Torso.
Source: Photograph courtesy of the author.

public accessibility to the edge of the quarry (Fernqvist 2010). The Green Party member on the Malmö district council also complained that the closeness of residential buildings to the edge of the quarry is likely to disturb the wildlife living on the cliffs (Börrefors 2008). Yet, the major political parties supported the scheme, bowing to the building companies and an alleged consumer demand (Fernqvist 2010).

One sure signal of Kalkbrottet as a place of gentrification is its relationship with Sweden's first gated community, entitled Victoria Park, which hugs the northern edge of the quarry. Victoria Park was developed as a "concept living premise" for people over 55 years of age (Olsson 2007). In combination with luxury condos, there is a spa, restaurant, and extensive services. A spokesperson for the complex stated that: "I really think that this is the future of living" (Kållberg 2008). Critics of the project, by contrast, claim the complex creates divisions between people and limits public access (Kållberg 2008, Alakoski 2009, Friberg and Lukic 2010, p. 40). The boundary-making surrounding Victoria Park reflects changing perceptions and practice on public access. When the complex was first built, boundary-making was focused on setting a limit between the nature reserve and the housing complex at the quarry's edge but left the complex open to public access. A decade later, boundary-setting focused on enclosing the green areas of Victoria Park which are now inaccessible to the general public. Meanwhile, the presence of the green-spotted toad at the private pool at Victoria Park has triggered a debate on whether the toads are public intruders and boundary-breakers.

The quarry figures prominently in the adverts for real estate that surround the quarry. Such advertisements tend to erase the past of the quarry while celebrating its scenic qualities and their focus for gentrification. One advert labels the housing at Kalkbrottet "a vision of nature and city life" and then continues:

Look closer at the spectacular nature and the unique limestone steppe. Imagine how it might be to live here already now. Dream away for a moment, lean back and familiarize yourself with the life on perhaps the most exciting place in Öresund. (HSB *et al.* 2012)

At the nearby cement factory, which closed in 1978, the gentrification process is now (2013) in full swing. In spite of several studies documenting its industrial heritage value, the cement factory is scheduled for almost complete eradication to make room for high-priced rental and condominium developments (Gustavsson 2009, Schlyter 2009). At the cement plant, the concern for industrial heritage is non-existent. The environmental gentri-fication process is less than subtle. At one of the old storage silos, one of the developers displayed a sign that depicted a cement mixer expounding a bouquet of flowers with the text: "So long concrete-Malmö: we are now transforming the old cement factory to a new and attractive part of the city. Welcome to Limhamn's location" (Figure 4). The sign symbolises the passing of an old and coming of a new era.

The gentrification of the neighbourhoods surrounding Kalkbrottet and the quarry itself is not only fuelled by real estate advertisements, but also a series of everyday practices. The residents of Victoria Park, for example, present themselves as people living in a collective who subscribe to common values in spite of living an exclusive and exclusionary existence (Friberg and Lukic 2010).

The physical presence of workers themselves is absent in the gentrified neighbourhoods that now surround the quarry. This is in part due to the early closure of the cement plant in 1978, leaving few workers still living today. It is also due to the quarry itself employing fewer and fewer workers over the years due to mechanisation (in 1901, 467 men worked in the quarry; in the early 1980s only 70 remained). But it is also a function of the gentri-fication of the neighbourhood. One blogger of a Limhamn environmental organisation,

Figure 4.   The old cement factory in Limhamn in June 2012.
Source: Photograph courtesy of the author.

commenting on Victoria Park, is less than subtle in commenting on the erasure of the working class people in Limhamn:

> Nowadays, there is so called 'seniors' living at Kalkbrottet, Victoria Park, though not for old and worn out cement – and quarry workers but for high society (sociechetten). I wonder if they send a thought to all those who worked there and got lung problems and typically coughed themselves through a short retirement. It doesn't seem that way when you read the advertise-ments with the supervisor there, a real lady of bling (glittermaja). One is offered a small apart-ment for lots of money. One has a spa and a lodge, as they are called nowadays, and everything costs money. It's the profits they live off. (Friberg n.d.)

One final reflection of the quarry as a source of exclusivity is its continued closure to the public. Public access to the quarry remains a contested issue within city politics. But so far, the local social democratic government has refused to offer the required funding to proceed with the development of public access to the quarry. Instead, it is primarily seen as a scenic amenity feature, a spectacularised place, part of the Örespectacle, and a commodity mar-keted in conjunction with the surrounding neighbourhood.

The environmental gentrification process in Malmö is confirmed by the city's record as a city promoting environmental, cultural, and economic rather than social goals (Salonen 2012). The city has recently, however, explored the social discrepancy in a commission on the "hidden Malmö", the part of the city that is poor and underserviced (Salonen 2012). Kalkbrottet may be a place where the environmental gentrification process is chal-lenged and where environmental efforts may serve to tackle Malmö's social deficit.

Kalkbrottet: a case against environmental gentrification?

What, if any, are the real and potential challenges to the environmental gentrification at Kalkbrottet? Is there anything socially and ecologically progressive about Kalkbrottet, something that runs counter to the environmental gentrification thesis, something that suggests that the quarry is more than an exclusive scenic amenity for surrounding real estate development (and only the occasional supervised visit for a selected number of the general public)? Is Kalkbrottet a mere spectacular post-industrial project which takes away resources from beneath the majority of ordinary everyday post-industrial areas and thereby masks "the rot beneath the glitter" (Harvey 2001, p. 364)? In Malmö, there are elements rooted in Malmö's social democratic past, the ecological dynamics in the quarry itself, and the people who speak and support these dynamics.

The quarry itself, I argue, is full of signs and events that open up discussion about the presence and possibilities of a different future. When closed in 1994, Kalkbrottet was var-iously considered for a housing complex, a golf course, an adventure park, a motocross track, a ski hill, and a dumpsite for the earth extracted in the building of traffic tunnels. The most extravagant prospect was a gentrification scheme called Scanstad, conceived by the architect Peter Broberg, which contained a housing and office complex that included an inhabited Italian-style bridge across a flooded quarry (Nohrborg n.d.).

But as the plans were considered at Kalkbrottet, a series of unexpected things happened. The abandoned quarry gradually became recognised as a refuge for a spontaneous invasion of flora and fauna, some of them endangered and rare for the local area and for Sweden more generally. Sixteen years later, in 2010, these flora and fauna formed the basis for the quarry's designation as a national nature reserve.

However, the mere presence of rare species was not enough to sanction the quarry as a nature reserve. The plants and animals had spokespeople who devised strategies to champion their cause. A group of city ecologists, with Mats Wirén as a central figure, emerged as the key proponents. Wirén was part of a specific division of the city adminis-tration, Gatukontoret, which was particularly active in support of the quarry as a nature pre-serve. The situation resembles the conception of Sweden as a strong administrative state, where administrators, bureaucrats, and certain individuals can yield considerable leverage against politicians. As Wirén recalls, in the efforts to designate Kalkbrottet as a nature reserve, he searched species' lists for different groups of rare organisms in the area. During visits to Kalkbrottet, he then looked for these flora and fauna (Wirén 2007). Some of the species he found, including the green-spotted toad (*Buffo viridis*) and kalkk-rassingen (Sw.; *Sisymbrium supinum*), proved decisive in gaining approval for the nature reserve designation. But other species were also strategically enrolled to play a role in the formation of the nature reserve. Collectively, they were neatly categorised, documented, and mapped in the material that formed the basis for successive successful applications for conservation status. In 2003, the City of Malmö designated Kalkbrottet as a future nature and recreational area in its Green Plan. And in 2007, the European Union accepted the eastern part as a Natura 2000 area, a protected site, as a result of its unique flora and fauna. These designations presented a serious obstacle to any future development of the site and eventually led to the all-encompassing national designation in 2010.

These achievements were hard won. At one point, HeidelbergCement forbid Wirén to visit the quarry and on another occasion shut him out of the deliberation surrounding the formation of Scanstad on account of the status of Kalkbrottet as private property. At a later stage, he returned from paternity leave to find a golf project well advanced in its plan-ning stage, a project he successfully fought to have reversed. The promoters of the nature reserve also had to fight hard to secure the funds to pump the water out of the quarry. Other similar quarries in the neighbouring area have been flooded and now constitute giant lakes. The ecologists at the City of Malmö, therefore, had to make a case for justifying the con-siderable expense of using pumps to keep the quarry dry.

Wirén offers a cautionary perspective on the built-up areas and the negative effects of gentrification that surround the quarry, musing that it could have been worse, given some of the strategies and efforts expended to sell and strike compromises with competing city offi-cials and developers to gain the current form of conservation of Kalkbrottet. In Malmö, Wirén avers, everything, including conservation efforts, has to be audacious ("fräckt") and saleable. In promoting Kalkbrottet as a nature reserve, the conservationist interests of the City had to make compromises in order to "save" anything at all. They also took every opportunity to publicise the quarry in the local media and in special publications. Wirén refers to a photographic essay in book form that appeared in two editions and that he circulated to politicians and the media, resulting in a huge impact in rallying support for the conservation of the quarry (Malmö Stad 2010a).

The transformation of the Limhamn quarry from an urban industrial to a green urban post-industrial landscape can at one level be seen as a success story against gentrification, a story of rare plants and animals finding a new home, and ecologists fighting and striking compromises with developers, city politicians, and bureaucrats to achieve the nature con-servation designation. It is not an unusual story, but one repeated in many city regions where declining industrial areas are subject to reclamation and restoration efforts of various kinds. In Malmö, it is the dominant story of Kalkbrottet, a story told by city ecol-ogists, and that now permeates and is celebrated in the popular media and public discourse.

The history of the quarry is often narrated as one where elements of surprise and nature's "own" agency are accepted. Surprising ecological elements stem back to at least the 1940s. In one part of the quarry, an old-growth woodlot of *sälg* (Sw.), goat willow (salix capria) still prospers and is classified as a rare floral community. At the same time, eel, *ål* (Sw.; *Anguilla anguilla*), was introduced in some of the deep dams of the quarry and they still survive and thrive. Managers also accept that the persistence of some rare species, which have a more recent history in the quarry, may not be ensured. After a flood in 2007, for example, the rare *kalkbräken* (Sw.; *Gymnocarpium robertianum*) is feared to have disappeared (Malmö Stad 2010b, p. 21).

The fauna of the quarry may also find new ways to arrange their lives. In one of the small dams of the quarry, for example, the heron, *hägern* (Sw.), has found small little islands to nest in place of their usual tree sites (Malmö Stad 2010c, p. 23). In another case, Wirén reports the movement of one of the giant eels from the deep dams in the western parts of the quarry to the eastern parts, where it helped itself to the rare green-spotted toads.

But by and large, the City managers have employed interventionist methods to design and police the ecologies at Kalkbrottet. These measures go against the fashionable "let nature takes its own turn" approach of some of the landscape architect firms consulted by the City. In order to control the flux and surprise in the quarry, the current conservation plan for the quarry is highly structured, with different zones emphasising different features, and different management strategies aimed at maintaining the quarry's unique ecological and geological character. Some of the interventions currently in place or planned in the quarry clearly reflect a human-guided aesthetic. These interventions include the suppression of specific species, such as the highly invasive *hagtorn*, hawthorn (*Crataegus*), which thrives in a limestone or karst landscape. Other techniques involve disturbing the ground to prevent the formation of soil that would change the unique landscape. Yet other measures include the control of the edible frog that preys on the rare green-spotted toad and the grey goose, which threatens to exhaust an already sparse vegetation cover. Managers also clean canals and other water bodies, remove falling stones from cliff bases, secure and restore existing industrial buildings and objects, implement security measures, and hope to establish walking paths and viewing sites, and building a visitors' centre (Malmö Stad 2010a, p. 4).

These overall measures suggest a highly controlled non-human environment that needs monitoring, manipulation, money, and policing. These measures serve something that is taken very seriously (though not mandated) by the planners: public access. The city ecologists have over time fostered a public affinity towards Kalkbrottet through frequent field excursions into the quarry. Tickets to such excursions are given out for free on a first come first service basis at City Hall. The demand is extremely high and the tickets are gone quickly. The field tours cover the quarry's unique natural and industrial heritage features.

There are now organised tours into the quarry on a regular basis but the ultimate goal is to provide such access on a more permanent basis. Future plans include different trails that both circle and descend into the quarry. There are plans for bathing in one of the dams and the use of a sauna. There are also spots for an amphitheatre and overnight sleeping. Another area of the quarry will be set aside for the exhibition of land art projects. These concerns for public access to "natural" and "recreational" sites have a long history in Malmö. Ribersborgstranden, the popular local beach, was constructed in the 1930s by truckloads of sand to create an open air leisure landscape to provide the working class with an opportunity for recreation and communing with nature (Ristilammi 1997; Figure 2). The sand at the Beach is continuously replenished as it erodes southward by water currents. And in Västra Hamnen, Malmö's old port area that has been transformed into a tony housing complex (anchored by the famous Turning Torso building), the waterfront is accessible and widely used by the public at large.

These plans and rhetoric are peasant society and social democratic legacies emphasising accessibility (*tillgänglighet*) to nature and social equity (*jämlikhet*) in the Malmö region, a point that also resonates with many Swedes' support of the notion of "a fair share of environmental space" (Isenhour 2011, p. 129, see also Pickett and Wilkinson 2010). In the province of Skåne, there is a tradition of using most space for cultural and economic activities, limiting access and the space for recreational pursuits in non-human natures, but there are prominent examples of projects where artificial natures have been constructed for such purposes. The Swedish traditions of *allemansrätten* and *friluftsliv* still matter. The goal of providing nature experiences to the broader public remains a goal into the present, at least in the official message of the City. In the *Green Plan for Malmö 2003* (Malmö Stad 2003) a couple of objectives function to advance this overall goal. For instance, the plan seeks "to create a variety of park, natural and recreation areas that together with specific recreation areas and green gardens provide for the 'green needs' of Malmö's population", and "to create a cohesive green network throughout the whole of Malmö with high accessibility".

But are there more ways in which Kalkbrottet could be conceived of as a social nature that accommodates more people and voices? There are certainly attempts to make the history of the quarry more visible in terms of its working class history. There are displays and photos at the quarry depicting workers and working conditions. At a City Hall exhibit, a special section contains some of the equipments used by the quarry workers and historical photos of their work. But such displays can freeze history in the past, gloss over the class struggles and consequences of a capitalist economy, and hide the social and environmental effects of the superquarries on the island of Gotland that have displaced the extraction at Kalkbrottet. In another context, James O'Connor (1998) has identified two seemingly distinct groups in the appreciation of an old but now naturalised quarry landscape in California. One group is interested in natural history while the other prefers industrial heritage history. At second blush, however, the groups have more common than different interests. Both groups come from the same social class of professional groups who focus their fascination on natural and industrial heritage at an isolated site rather than viewing its larger social and environmental context.

Doron (2000, p. 252) provides a deeper rationale for symbolically maintaining these landscapes as reminders of the exploitative history of capitalism. He writes:

> The effects of post-industrialism, the passing of time, wars, the nature of capitalism, and parsimonious speculation are, of course, the things that produced these places. ... abolishing these places by re-planting, redeveloping, revitalizing and Renaissanciation, will not delete the systems that produced these 'voids' and waste. These strategies and their methods are part of the same economic, social, political and planning systems that created these places from the beginning. Replanning, redeveloping, revitalizing and Renaissanciation will simply erase the evidence of the crime. Furthermore, it will exterminate the victims who found refuge in the waste that these systems, for the own sake, deliberately produced.

There may be other ways to engage different voices at Kalkbrottet. In the past, "illegitimate" users and actors included scrap metal collectors, adventure seekers, and temporary residents. The signs of graffiti artists are now ubiquitous on the old industrial structures in the quarry and then often posted as You Tube videos and photos on the World Wide Web (Figure 5). Another segment of trespassers falls into the category of urban explorers who visit, walk, and photograph the abandoned industrial buildings without permission and then post them on the web. One site in particular, Tillträde Förbjudet (2012), receives considerable traffic and often features in the local and national newspapers.

Figure 5.   Konsthallen, the art gallery, in one of the old industrial buildings in Kalkbottet. June 2012. Source: Photograph courtesy of the author.

Wirén speaks of these trespassers as often practising risky behaviour and disturbing wildlife and constituting a liability for the city. These are legitimate concerns. But there are few efforts by city managers to incorporate these informal visitors into the history or official plans of the quarry. In contrast to such exclusion, Jacques (2011) considers such users as "improvisers" who can exhibit "vitality and intensity" within a "demotic or informal area" of the city. Post-industrial landscapes are therefore never dead zones, they contain living histories that need to be remembered and problematised.

There are other parties who could be more effectively engaged in Kalkbrottet. In particular, Kalkbrottet could be a place to mediate and diffuse the tensions between ethnic Swedes and recent immigrants. Wirén describes a group of Somali immigrant refugee women who visited the quarry and were reminded of home in the warm continental climate and the sparsely vegetated limestone plains. For them, the quarry constituted a refuge and a place of consolation in a struggle to find a place in Swedish society. Kalkbrottet may in fact constitute what Nouwen (1972, p. 84) refers to as a wounded healer, "a deep incision on the surface of our existence" that can serve as "an inexhaustible source of beauty and self-understanding". Other city-sponsored projects confirm the value of nature outings for the well-being of new immigrants (Lisberg Jensen 2009). Yet such initiatives do not appear to be extensive. There are no affirmative action measures to promote the presence of new immigrants at Kalkbrottet. When visiting the quarry itself, walking the neighbourhoods surrounding it, and talking with stakeholders for five days, I did not encounter any new immigrants.

There are, then, ways in which Kalkbrottet could be a place that brings together difference that is so prevalent in the city today. One of the City agencies expresses this goal as follows:

We want a city where all people have equal value and where diversity is considered an asset . . . We want a city without fear of the 'other' and void of discrimination and racism. We have to see both the differences and similarities among us as natural and self-evident. To make this possible, we need more encounters between people of different backgrounds, cultures and religions at work and at home. (Malmö, Integrationsrådet, quoted in Lundén 2010, p. 3)

In a city with a population of 300,000, an efficient public transit system, and an extensive network of bike lanes, Kalkbrottet is readily accessible to most residents in Malmö. However, it is not clear how the gentrification of the surroundings of Kalkbrottet and Sweden's first gated community contributes to the fulfilment of this vision.

Conclusion

One lesson from this case study suggests that Kalkbrottet follows the classic pattern of environmental gentrification, a process that is tied up with the neoliberalising and competitive city, the city that pursues the audacious or ostentatious, something that stands out, something spectacular, and something that builds a city that attracts business and cultural elites who are not only producers but consumers of emblems of efficient commuting (accessible public transit rather than gridlock car traffic), condominium living, (st)architecture, security, good shopping, world-class cultural events and happenings, and, of course, a green environment. Kalkbrottet itself and the rare flora and fauna that have ventured into it have become neighbours with wealthy, white and well-established Swedes, professionals and retirees, the 55 + crowd, who peer out the window in front of Sweden's "Grand Canyon". From this perspective, Kalkbrottet is primarily a scenic amenity, part of the Örespectacle, where public access, except for the occasional supervised tour, does not play a prominent role.

Yet the events at Kalkbrottet can also be seen from a different perspective, one of a circumscribed neoliberalism containing surprise elements of nature, activism of city ecologists, and traditional and social democratic ideals promoting accessibility and equity. City ecologists have resisted the environmental gentrification process and developers and their supporters in the City government have had to temper their image of the quarry as a housing development, at one time inside the quarry and at other times close to the quarry edge. And though the quarry itself now remains closed to the public because of a lack of political will to fund it, its present plans contain generous accommodation for public access to pursue nature walks and recreation in the quarry. From this perspective, the public opening of the quarry remains an important goal and a political and contested issue.

The city ecologists have thus played a prominent position, but in a different role from the promoters of nature conservation elsewhere in Sweden, for example, the central state taking the initiative to write a national nature park narrative or a diverse urban movement of civic society organisations playing a prominent role in advocating for a regional park in Stockholm (Mels 2002, Ernstson and Sörlin 2009). To be sure, the flora and fauna in the quarry, besides being strategically enrolled by the city ecologists, have also spoken up through their presence, beauty, and unpredictable ways.

Yet the resistance towards environmental gentrification at Kalkbrottet is itself circumscribed. The signs of former quarry workers and industrial processes in the quarry, while present and displayed, could be more forcefully asserted as a voice in the history of the site. There are other untapped potentials in promoting a broader access to the quarry. There could be more attempts to meet the needs of "rogue" elements who now enter the quarry illegally, and whose presence is the most visible through graffiti art. There could likely be more efforts to accommodate access to Malmö's growing immigrant population. In one of his songs, long-time Malmö resident and singer and activist Mikael Wiehe sings "Vad bryr sig kärleken om gränser?", "What does love care about borders?" which is about a love story between an immigrant man and a local woman. Kalkbrottet has the potential to speak to the same message by constructing a meeting place for "others": by recreating the physical geographies and climates where new immigrants come from, by celebrating an industrial and working class history, by harbouring a unique flora and fauna, by providing a home for the outcasts of society, and by accommodating and bringing together these different cultures and activities.

References

Alakoski, S., 2009. *Överlever folkhemmet när de rika isolerar sig? Aftonbladet* [online], 27 September. Available from: http://www.aftonbladet.se/debatt/debattamnen/samhalle/article1201 2782.ab [Accessed 10 May 2012].

Baeten, G., n.d. *Producing the neo-liberal city: Hyllievång, Malmö* [online]. Available from: http://www.uib.no/ngm/content/Papers/Critical%20geography.pdf [Accessed 23 April 2012].

Baeten, G., 2011. *Swedish neoliberal cities in 'Post-neoliberal' times* [online]. Neoliberalism in Post-Welfare State Workshop. Available from: http://www.ubc.ca/okanagan/cssej/publishing/cssejpress/Neoliberalism_and_Post-Welfare_Nordic_States_Workshop_Presentations/Dr__Guy_Baeten_-_Swedish_neoliberal_cities_in__post-_neoliberal__times.html [Accessed 12 May 2012].

Baeten, G., 2012. Normalizing neoliberal planning: the case of Malmö. *In*: T. Taşan-Kok and G. Baeten, eds. *Contradictions of neoliberal planning: cities, policies and politics*. Dordrecht: Springer, 21–42.

Bélanger, A., 2000. Sports venues and the spectacularization of urban spaces in North America: the case of the molson centre in Montreal. *International Review for the Sociology of Sport*, 35 (3), 378–397.

Börrefors, Bo. 2008. *Klart for Bostäder vid Kalkbrottet. Skånskan* [online], 24 April. Available from: http://www.skanskan.se/article/20080424/NYHETER/307237767/-/klart-for-bostader-vid-kalkbrottet [Accessed 12 June 2012].

Broudehoux, A., 2010. Images of power: architectures of the integrated spectacle at the Beijing Olympics. *Journal of Architectural Education*, 63 (2), 52–62.

Bryson, J., 2012. Brownfield gentrification: redevelopment planning and environmental justice in Spokane, Washington. *Environmental Justice*, 5, 26–31.

Checker, M., 2011. Wiped out by the "greenwave": environmental gentrification and the paradoxical politics of urban sustainability. *City & Society*, 32 (2), 210–229.

Clark, E. and Johnson, K., 2009. Circumventing circumscribed neoliberalism: the 'system switch' in Swedish housing. *In*: S. Glynn, ed. *Where the other half lives: lower income housing in a neoliberal world*. London: Pluto Press, 173–194.

Cunningham, H., 2012. Gated ecologies and 'possible urban worlds': from the global city to the natural city. *In*: I. Leman Stefanovic and S. Bede Scharper, eds. *The natural city: re-visioning the built environment*. Toronto: University of Toronto Press, 149–160.

Curran, W. and Hamilton, T., 2012. Just green enough: contesting environmental gentrification in Greenpoint, Brooklyn. *Local Environment*, 9 (1), 1027–1042.

Debord, G., 2006. *Society of the spectacle*. London: Rebel Press.

Dooling, S., 2009. Ecological gentrification: a research agenda exploring justice in the city. *International Journal of Urban and Regional Research*, 33, 621–639.

Doron, G., 2000. The dead zone and the architecture of transgression. *City: Analysis of Urban Trends, Culture, Theory, Policy, Action*, 4 (2), 247–263.

Ernstson, H. and Sörlin, S., 2009. Weaving protective stories: connective practices to articulate holistic values in the Stockholm national park. *Environment and Planning A*, 41, 1460–1479.

Fernqvist, F., 2010. *Öster i Kalkbrottet i Limhamn* [online]. Available from: http://fernqvist.blogspot.com/2010/03/oster-om-kalkbrottet-i-limhamn.html [Accessed 4 June 2012].

Foster, J., 2005. The don valley brickworks: whose restoration? Whose space? *The Journal of Urban Design*, 10, 331–352.

Foster, J., 2010. Off track, in nature: constructing ecology on old rail lines in Paris and New York. *Nature and Culture*, 9 (3), 316–337.

Friberg, L., n.d. *En Annons, Ett Hus, och Många Minnen* [online]. Available from: http://www. limhamn.org/modules/tinyd0/index.php?id=32 [Accessed 15 May 2012].

Friberg, E. and Lukic, A., 2010. *Gemenskap är det Svarta – En studie av Victoria Park som ett Gated Community*. C-Uppsats. Malmö Högskola.

Gelter, H., 2000. *Friluftsliv*: the Scandinavian philosophy of outdoor life. *Canadian Journal of Environmental Education*, 5, 77–92.

Groth, J. and Corijn, E., 2005. Reclaiming urbanity: indeterminate spaces, informal actors and urban agenda setting. *Urban Studies*, 42 (3), 503–526.

Gustavsson, O., 2009. *Landskapsurbanism i en Postindustriell kontext* [online]. Alnarp: Sveriges Lantbruksuniversitet. Available from: http://stud.epsilon.slu.se/488/1/gustavsson_o_090914.pdf

Harvey, D., 2001. *Spaces of capital: towards a critical geography*. New York: Routledge.

Harvey, D., 2005. *A brief history of neoliberalism*. Oxford: Oxford University Press.

Hedin, K. *et al.*, 2012. Neoliberalization of housing in Sweden: gentrification, filtering, and social polarization. *Annals of the Association of American Geographers*, 102 (2), 443–463.

Herrington, S., 2009. *On landscapes*. New York: Routledge.

Higgs, E., 2003. *Nature by design: people, natural processes and ecological restoration*. Cambridge: The MIT Press.

Hobbs, P. and Sandilands, C., 2013. Queen's park and other stories: Toronto's queer ecologies. *In*: L.A. Sandberg *et al.*, ed. *Urban explorations: environmental histories of the Toronto region*. Hamilton, ON: L.R. Wilson Institute of Canadian History, 72–94.

HSB, *Sjælsø Gruppen and Veidekke* [online], 2012. Available from: http://www.kalkbrottet.nu/? page=page47ee23f199803 [Accessed 24 March 2012].

Igoe, J., 2011. The spectacle of nature in the global economy of appearances: anthropological engagements with the spectacular mediations of transnational conservation. *Critique of Anthropology*, 30 (4), 375–397.

Isenhour, C., 2011. How the grass became greener in the city: on urban imaginings and practices of sustainable living in Sweden. *City & Society*, 23 (2), 117–134.

Ivarsson, D., 2007. *Det Perfekta Brottet Ligger i Limhamn* [online]. *Sydsvenskan*, 5 April. Available from: http://www.sydsvenskan.se/malmo/det-perfekta-brottet-ligger-i-limhamn/ [Accessed 1 April 2012].

Jacques, P., 2011. Urban improvisations: the profanatory tactics of spectacularized places. *Critical Studies in Improvisation* [online], 7 (11). Available from: http://www.criticalimprov.com/ article/viewArticle/1390/2023 [Accessed 4 November 2012].

Jamison, A., 2008. Greening the city: urban environmentalism from Mumford to Malmö. *In*: M. Hård and T. Misa, eds. *Urban machinery: inside modern European cities*. Cambridge: The MIT Press, 281–298.

Kållberg, M., 2008. *Lyxboende – med vakter* [online]. *Expressen*, 4 June. Available at: http://www. expressen.se/levabo/1.1186188/lyxboende-med-vakter [Accessed 1 May 2012].

Kirkwood, N., ed., 2001. *Manufactured sites: rethinking the post-industrial landscape*. London: Spon Press.

Koglin, T., 2008. *Neo-liberalising the countryside of Skåne – the golf course development in Bara*. Master's thesis. München: Grin Verlag für akademische Texte.

Latour, B., 2005. *Reassembling the social: an introduction to actor-network-theory*. Oxford: Oxford University Press.

Lefebvre, H., 1996. *Writings on cities*. E. Kofman and E. Lebas, trans. and ed. Oxford: Blackwell.

Lehrer, U., 2003. The spectacularization of the building process: Berlin, Potzdamer Platz. *Genre*, XXXVI, 383–404.

Lindvall, J. and Rothstein, B., 2006. Sweden: the fall of the strong state. *Scandinavian Political Studies*, 29, 47–63.

Lisberg Jensen, E., 2009, "Gud har varit givmild här": Friluftsliv, etnicitet och ekologisk läskunnighet i Arrie. *In*: P. Hillbur, ed. *Närnaturens Mångfald*. Malmö: Malmö University Publications in Urban Studies (MAPIUS), Vol. 3, 197–219.

Lundén, J., 2010. *Malmö – Möten och Mångfald? En studie av bostadssegregationen i Malmö med fokus på Västra Innerstaden och Limhamn-Bunkeflo*. C-Uppsats. Malmö Högskola.

Lund Hansen, A., 2010. Forty years of system change: lessons from the free city of Christiania. *In*: L.A. Sandberg and T. Sandberg, eds. *Climate change – who's carrying the burden? The chilly climates of the global environmental dilemma.* Ottawa: Canadian Centre for Policy Alternatives, 217–230.

Lundin, J., 2006. *Näten på Limhamn: social relationer in ett lokalsamhälle, 1870–1914.* Lund: Sekel.

Malmö Stad, 2003. *Green plan for Malmö 2003 summary* [online]. Available from: http://www.malmo.se/download/18.1c002f7b12a6486c372800012092/Gronplan_summary+ENG.pdf [Accessed 1 June 2012].

Malmö Stad, 2010a. *Bildande av Naturreservat Limhalms Kalkbrott* [online]. Beslut. Available from: http://www.malmo.se/download/18.2d03134212cf2b7c00b800039749/Naturreservat+Limhamns+kalkbrott_Beslut+2010.pdf [Accessed 15 May 2012].

Malmö Stad, 2010b. *Naturreservat Limhamns Kalkbrott. Skötselplan* [online]. Available from: http://www.malmo.se/download/18.2d03134212cf2b7c00b800039763/Naturreservat+Limhamns+kalkbrott_Skotselplan+2010.pdf [Accessed 16 May 2010].

Malmö Stad, 2010c. *Limhamns Kalkbrott – En Unik Plats I Malmö.* 2nd ed. Malmö Stad: Gatukontoret.

Malmö Stad, 2012. *Kalkbrottet, Limhamn* [online]. Available from: http://www.kalkbrottet.nu/?page=mgc [Accessed 12 June 2012].

Mels, T., 2002. Nature, home and scenery: the official spatialities of Swedish national parks. *Environment and Planning D: Society and Space*, 20, 35–154.

Newman, A., 2011. Contested ecologies: environmental activism and urban space in immigrant Paris. *City & Society*, 23 (2), 192–209.

Nohrborg, K., n.d. *En Stad Växer Upp I Limhamns Kalkbrott* [online]. Available from: http://www.limhamnsmuseum.nu/PDF%20filer/KN.PDF [Accessed 12 April 2013].

Nouwen, H., 1972. *The wounded healer: ministry in contemporary society.* New York: Doubleday.

O'Connor, J., 1998. *Natural causes: essays in ecological Marxism.* New York: Guildford Press.

Olsson, E., 2007. *Kalkbrottet lönar sig inte. Kvällsposten* [online], 21 March. Available from: http://www.expressen.se/kvp/kalkbrott-lonar-sig-inte/ [Accessed 3 April 2012].

Pickett, K. and Wilkinson, R., 2010. *The spirit level: why greater equality makes societies stronger.* New York: Bloomsbury Press.

Qviström, M. and Saltzman, K., 2006. Exploring landscape dynamics at the edge of the city: spatial plans and everyday spaces at the inner urban fringe in Malmö, Sweden. *Landscape Research*, 31 (1), 21–41.

Ristilammi, P., 1997. Strandvisioner: om skapelsen av ett modernt stadsrum. *In*: K. Saltzman and B. Svensson, eds. *Moderna landskap: Identifikation och tradition i vardagen.* Stockholm: Natur och Kultur, 91–110.

Salonen, T., 2012. *Befolkningsrörelser, försörjningsvillkor och bostadssegraegation: en sociodynamisk analys av Malmö* [online]. Malmö: Malmö Stad. Available from: http://www.malmo.se/download/18.d8bc6b31373089f7d9800059627/Det+dolda+Malmö_Tapio+Salonen_Malmökommissionen_final.pdf

Schlyter, O., 2009. *Cementfabriken och Bangården i Limhamn* [online]. Malmö: Malmö Museer. Available from: http://www.malmo.se/download/18.af27481124e354c8f1800042629/Rapport_2009_018_low.pdf

Smitt, R., 2007. *Malmöföretagen – Förr och Nu.* Lund: Wallin och Dalholm.

Storm, A., 2008. *Hope and Rust: Reinterpreting the Industrial Place in the late 20th Century* [online]. Stockholm: Division of History of Science and Technology, Royal Institute of Technology, KTH. Available from: kth.diva-portal.org/smash/get/diva2:13187/FULLTEXT01 [Accessed 28 May 2012].

Svensson, P., 2010. *En Stad av Grus. Sydsvenskan* [online], 5 September. Available from: http://www.sydsvenskan.se/kultur–nojen/en-stad-av-grus/ [Accessed 5 May 2012].

Tillträde Förbjudet. 2012. Available from: http://www.tilltradeforbjudet.se/ [Accessed 21 May 2012].

Wirén, M., 2007. Kalkbräken in Limhamns Kalkbrott – en ny och isolerad fyndplats. *Svensk Botanisk Tidskrift*, 101 (3–4), 237–238.

Buried localities: archaeological exploration of a Toronto dump and wilderness refuge

Heidy Schopf[a] and Jennifer Foster[b]

[a]Cultural Heritage Specialist, Archaeological Services Inc., Toronto, ON, Canada; [b]Faculty of Environmental Studies, York University, Toronto, ON, Canada

The Leslie Street Spit is best known as an urban wilderness refuge but it has a fascinating, although obscure, social history. Archaeological methods are used here to uncover the material associations between the Leslie Street Spit and the City of Toronto. This approach reveals that the Spit reflects the past planning practice and creative destruction of the city. The Spit is found to contain artifacts of the past such as domestic items and rubble that resulted from slum clearing practices of the 1960s and development-driven planning practice of the 1980s. In its present state, the Leslie Street Spit acts as the romanticised ruins of the City of Toronto, composed of the material elements of the city that were discarded so that new and "up-to-date" forms of architecture could take their place.

The Leslie Street Spit is a 5-km manufactured peninsula that extends from Toronto's old industrial lands into Lake Ontario. It is composed of construction debris, and is a world-class birding site that hosts rare and endangered species. This paper examines the discarded and buried artifacts that underlie this celebrated post-industrial landscape and discovers that these artifacts, when contextualised through archaeological research methods, tell a story that previously had been obscured. It is a story about urban development processes, the destruction of the built heritage of Toronto, displacement of poor communities that got in the way of modernist ideals, and the ability of nature to transform industrial space into romanticised ruins. In this case, the act of memory suppression is performed in two specific ways: by omission and misrepresentation of the contents of refuse transported to the Spit from policy and planning records, and by the power of nature to distract civic attention from critical awareness of what has actually been destroyed.

The Toronto Harbour Commissioners (now the Toronto Port Authority) began dumping construction rubble and lake dredgeate on the edge of Lake Ontario in 1959. The dumping continues today, and the Leslie Street Spit is now over 500 hectares in size (Toronto Region and Conservation Authority 2010). Over time, this landmass has been gradually colonised by seeds and plant matter dispersed by wind, birds, water, and deposited material. Through this process, and in concert with ecological restoration efforts, the Spit boasts wildflower meadows, cottonwood forests, coastal marshes, cobble beaches, and sand dunes (Yokohari and Amati 2005, p. 55). Over 390 plant species and 290 animal species have colonised the Leslie Street Spit (Yokohari and Amati 2005, p. 55). Described as an "inspired accident" by Fulford (1995), the Leslie Street Spit is celebrated as a symbol of wilderness in the city given its impressive array of wildlife and rich ecology (Foster and Sandberg 2004, p. 191). Even more remarkable, the Spit's rich ecological profile coincides with active construction waste dumping during the week, when recreational park users are prohibited. Currently accepted materials at the Spit include unreinforced concrete, broken concrete, brick, ceramic tiles, and clean porcelain materials (Toronto Port Authority 2012). These materials are considered clean fill and are allowed to be dumped at the Spit during the week when recreational use of the park is restricted. Despite limited access, the Spit is fiercely protected from development by a highly organised network of concerned citizens. Much of the appeal of the Spit lies in popular appreciation for what is perceived as an untamed, sublime, and feral aesthetic, where nature is able to heal the scars of industrialisation. As Foster (2007) explains, "It juxtaposes a degraded and discarded city with fertile and vigorous ecology, a place where nature has colonized the post-industrial urban spoils" (p. 122).

While the diverse wildlife and rich ecology have become the defining features of the Spit, it also has a fascinating, although lesser known, social history that is embedded in the rubble that forms the material foundation of the landscape. Reconstructing this history is like a puzzle where we know the outcome of over 60 years of dumping into Lake Ontario, but not the role that the chunks of concrete and brick played in the City of Toronto before their deposition at the Spit. Reconnecting the rubble of the Spit and the destroyed elements of the city helps build insight into the history of what was destroyed, who was displaced, and what was constructed to fill the void created by the demolition of Toronto's built heritage. Understanding the history of this landscape and giving context to the rubble of the Leslie Street Spit is the main intent of this research.

Archaeological research and practice has not yet been explored in environmental justice scholarship. In many cases, and especially in instances where communities are completely displaced from urban landscapes, archaeological research may provide the only entry point for understanding previous landscape relationships. Sometimes the only traces of displaced livelihoods are buried in dumps. As documented by Rathje and Murphy, using archaeological methods to study dumps and refuse sites is a well-established practice and can lend insight to social conditions that are otherwise inaccessible through other modes of research (2001). In this case, the appeal of the Spit hinges on a popular understanding of the space as composed of construction waste (anonymous and benign materials) and as opposed to personal artifacts (evoking human subjects dissociated from their households). Using archaeological research methods allows us to challenge this notion and better appreciate the underlying nature of the landscape.

This research aims to give a general historic context to the Leslie Street Spit by coupling sections of the Leslie Street Spit with contemporary planning practice and urban development projects in the City of Toronto. In particular, the activities of the City during the years 1964 and 1980 are studied and linked to specific sections of the Leslie Street Spit. The exercise reveals that the Spit is linked to slum clearance and urban renewal initiatives of the 1960s and the development-driven construction boom of the 1980s. Most surprisingly, this research finds that the 1960s' deposits contain high levels of personal artifacts, suggesting that whole households were demolished and dumped at the Spit. This discovery challenges the claim that the Leslie Street Spit is solely composed of "clean fill" and rather suggests that early dumping activities included food waste, personal items, and household debris in addition to construction rubble. The key findings of this research illustrate that the Leslie Street Spit is not just a landscape defined by its wilderness, but is also a landscape defined by the development, destruction, and renewal of the built form of the city.

Archaeological research, displaced context, and creative destruction

The rubble that forms the Leslie Street Spit has an obvious relationship with the city, but this relationship is obscure since it has no specific ties to its former use and historical context. In archaeological theory and practice, it is crucially important to understand the context of an artifact in order to reconstruct past human activity, as it is the associations between artifacts that give a site meaning, not just the artifacts themselves. A great amount of information about the society that produced a site can be determined by examining the associations between artifacts (Renfrew and Bahn 2004, p. 55). In this sense, the Leslie Street Spit is a landscape that is entirely out of context. All of the rubble originated kilometres away and was transported to the Spit to be mixed with rubble that may or may not have anything to do with its former use in the city. While the majority of material found at the Spit cannot be linked to the city, some individual pieces, such as marked brick and the makers mark on a teacup, can be traced and dated. This is further aided by the fact that the Leslie Street Spit was constructed in a linear manner with cumulative dumping radiating from the headlands into Lake Ontario so that a general time period of material accumulation can be assigned to each section of the landscape (Figure 1). The chronological deposition of the materials at the Spit is key to assigning historical context to the rubble.

A central concern for this research is understanding the particular acts of demolition and planning trends within the City during the years 1964 and 1980. Studying the urban development activities of these two years gives clues to the possible origins of the rubble at the Spit. To better understand these processes, David Harvey's concept of "creative destruction" is used to articulate of the processes of urban construction and demolition that are evident at the Spit. Examining these discarded pieces of the built form of Toronto through a lens of creative destruction helps us to understand the political context that shaped Toronto and, in turn, the Leslie Street Spit. Harvey writes,

> One of the myths of modernity is that it constitutes a radical break with the past. The break is supposedly of such an order as to make it possible to see the world as a tabula rasa, upon which the new can be inscribed without reference to the past, or, if the past gets in the way, through its obliteration. Modernity is, therefore, always about "creative destruction," be it of the gentle and democratic, or the revolutionary, traumatic, and authoritarian kind. (Harvey 2006, p. 1)

In other words, elements of the older or vernacular built form of a city need to be destroyed in order for new architecture to overlay the city's identity. This process is used to signify that a city is up-to-date and willing to evolve with the times, and can be used to justify socio-political projects aimed at enhancing municipal prosperity by facilitating capital investment and removing obstacles to wealth accumulation such as working class or slum neighbourhoods in prime real estate.

Following Harvey's lead, the rubble of the Leslie Street Spit can be interpreted as the elements of the city that were destroyed to accommodate creation of new architectural forms that resulted from the process of creative destruction. Harvey's analysis of creative destruction focuses principally on neoliberal economic transitions, where states welcome

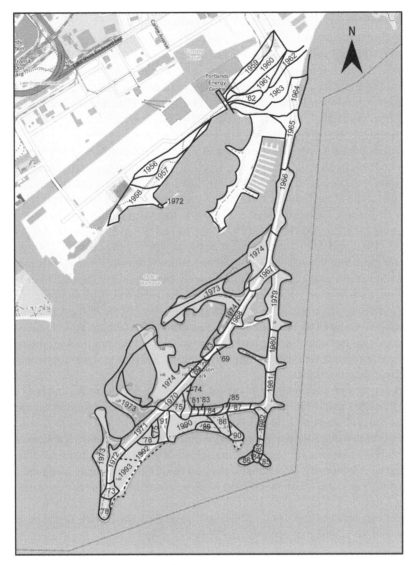

Figure 1.   Layout of the Leslie Street Spit.
Source: Base Map (Open Street Map 2013); Yearly divisions based on information from the Tommy Thompson Park Master Plan & Environmental Assessment (MTRCA 1992, p. 20).

new development by shifting to market-driven policies. We situate the physical destruction of entire neighbourhoods as a central element of Toronto's economic reform.
Zukin (1991) illuminates the impact of market culture on sense of place, and in this
case the process of accumulation and annihilation relates directly to replacement of
working class homes and the nineteenth-century buildings with some of Toronto's most
iconic modernist structures, ushering in a new relationship with development and developers. By linking creative destruction with the analysis of the Leslie Street Spit, we argue
that archaeological methods may offer important insight into contemporary environmental
justice concerns.

Research methods

Archaeological methods are here conjoined with research methodologies such as analysis of policy and planning documents, media materials and other records, archival research, and site analysis. Archival research is a key source of information in this study since much of the information about the origins of brick and aggregate activity in Toronto is not documented in a scholarly fashion. Rather, this information lies in old building reports, demolition reports, brick catalogues, business correspondence, historical photographs, maps, and indexes. Archival photographs and historic maps, in particular, are instrumental in understanding historic built forms of the city and in identifying individual structures and areas that have undergone dramatic change as a result of the process of creative destruction.

Fieldwork was conducted from May to July 2010 in order to identify traceable pieces of rubble at the Leslie Street Spit. Two sections of the landscape were selected for study based on the 1992 Tommy Thompson Master Plan from the Metropolitan Toronto and Region Conservation Authority (MTRCA) (1992). The MTRCA report indicates that the years 1964, 1968, 1980, and 1981 were the most active in terms of the disposal of construction material. Of these four years, 1964 and 1980 proved ideal for visual survey since they both are zones of high disposal activity and are currently exposed along the eastern edge of the Spit. In order to determine the current locations of the 1964 and 1980 dumping zones, the MTRCA 1992 map was overlaid on a current aerial imagery of the Leslie Street Spit. Once the 1992 map was placed on the current map, four Global Positioning System (GPS) points were selected that represent the north/south borders of the 1964 and 1980 dumping zones (Figure 2). These points were then programmed into a GPS unit to signal the boundaries of the respective dumping zones.

Visual survey and photographic materials are used to document the material composition of the 1964 and 1980 dumping zones. Both zones were identified through GPS points, marked with flagging tape and then surveyed. Visual survey consisted of scanning the exposed, eastern edge of the 1964 and 1980 zones for any marked brick and/or concrete, and any objects of interest (for example, household debris, personal items, or other traceable artifacts). Additionally, photographs were taken of the stratigraphy and general composition of the study areas. Follow-up field work was conducted to trace any marked brick back to its source locations, the results of which form the basis of a separate study analysing the connections between the Leslie Street Spit, aggregate sites, brick factories, regional buildings, and the rubble that comprise the bulk of the Spit.

Archaeological findings

The 1964 deposition zone at the Leslie Street Spit is located on the headlands, approximately 325 m south-southeast of the foot of Leslie Street and Unwin Avenue. The rubble found along the eastern shore of the 1964 zone is not uniform and seems to go through a gradual change in material composition starting from the northern end and extending south towards the 1965 zone. The northern-most edge of the zone has fully exposed stratigraphy, heavily worn brick and concrete, and is mostly colonised by vegetation. The southern-most part of the 1964 shore is mainly composed of large, heavily worn concrete pieces (see Figures 3 and 4, for examples, of north and south boundaries).

It is likely that the large concrete pieces in the southern portion of the 1964 zone were deposited after 1964 in order to protect the eastern shore from erosion. This practice started in the mid-1970s when the Toronto Harbour Commission (THC) required haulers to separate rubble from earth fill and use larger aggregate materials on the exposed outer face of the

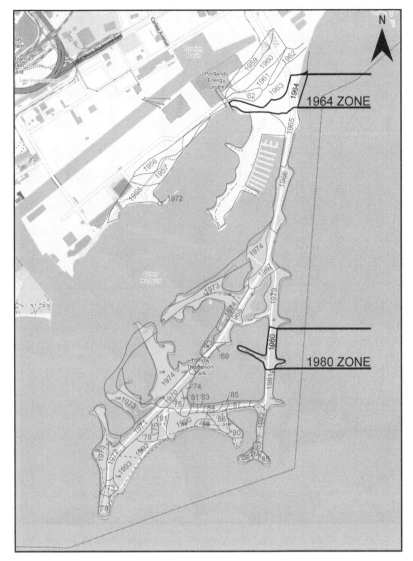

Figure 2.   Locations of the 1964 and 1980 zones at the Leslie Street Spit.
Source: Base Map (Open Street Map 2013); Yearly divisions based on information from the Tommy
Thompson Park Master Plan & Environmental Assessment (MTRCA 1992, p. 20).

Spit to guard against erosion (Ontario Ministry of the Environment 1982, p. 4). In light of
this, the exposed stratigraphy at the northern end of the 1964 zone reveals the most about
the dumping activities during this year since it can confidently be assigned to this year only.
Another interesting feature of the 1964 zone is the vegetation that dominates the landscape.
With the exception of the exposed eastern shore, there is little exposed rubble on the surface
of this zone since the bulk of this area is covered by grasses, trees, wildflowers, and other
forms of wildlife have colonised the Spit over time.

   One of the most interesting features of the 1964 zone is the exposed stratigraphic profile
near the northern end of the zone, since this stratigraphy demonstrates the loose (if non-
existent) dumping controls during this time period. High levels of household debris and

Figure 3.   South end of the 1964 zone. The concrete pieces were likely deposited after 1964 to protect the shore from erosion.

personal items protrude from the eroding wall, including teacups, bits of glass, medicine bottles, plates, diapers, electrical wire, rusted metal, eye glasses, toothpaste tubes, and even food waste (Figures 5 and 6).

The collection of material at this location on the Spit suggests that full houses with belongings still inside were demolished, compacted, and then dumped in Lake Ontario. Finding household debris, personal items, and food waste does not support the claim that the Spit is only composed of "clean fill". The loose standards of early dumping at the Spit are explained in the Ministry of the Environment's (MOE 1982) report, *Lakefill Quality Study: Leslie Street Spit*. Though household materials remain unacknowledged, the report reads, "Originally the quality of the fill materials was not of concern and, at the earliest stages of the lake filling, the materials were comprised mostly of excavated earth, construction rubble, dredge spoils, and miscellaneous solid waste" (MOE 1982,

Figure 4.   North end of the 1964 zone. The stratigraphy is compacted and can be assigned to a specific year of deposition with confidence.

p. 4). It was not until 1979 that any quality control system was put in place by the THC and even then, the only requirement was that trucks with fill materials had to pass a visual and olfactory inspection (MOE 1982, p. 4). In other words, the material just had to look and smell passable in order to be deemed suitable for dumping at the Spit. These controls, however, were not satisfactory in restricting contaminated materials from being dumped into Lake Ontario. It reads,

> Inspections by the W.D. Wilkins and Associates field staff of the trucked material dumped onto the Spit revealed that the visual and/or olfactory inspections of the THC were not effective in preventing contaminated materials from gaining access to the Leslie Street Spit and possibly Lake Ontario. (MOE 1982, p. 9)

Figure 5.   Bottom of a teacup that is compacted in the 1964 rubble.

Figure 6.   A sawn long bone compacted in the 1964 rubble. The bone is likely from an ungulate (e.g. cattle or pig).

This strongly suggests that the City was not careful in tracking the source and composition of the rubble being dumped at the Leslie Street Spit during the 1960s.

The bricks found in the 1964 zone display a variety of markings which give clues to the geographic locations of the source material and the companies that manufactured the brick.

These markings include: Domtar, Don, Don Valley, J.Price, Milton, Ontario P. B. Co, and Bessemer Block (Youngstown, Ohio). It should be noted that with the exception of Bessemer Block, which originates in Ohio, all these brick makers are local Ontario brick manufacturers with production facilities in Ontario.

The 1980 zone is located approximately 2.6 km south of the foot of Leslie Street and Unwin Street. The 1980 zone presents a very different landscape from the 1964 zone since there is little vegetation, and the construction rubble is plainly visible on the surface. The type of material found in the 1980 zone is also much more uniform and organised, suggesting that dumping had become standardised and regulated by this time. There is even a logic to the deposition of the materials. There is a small brick beach made up of worn brick surrounded by banks of concrete rubble that serve to divide the years of deposition and provide harbour-like conditions for any wildlife that may establish there (Figures 7 and 8). No personal or household artifacts were found in this section of the Spit. The rubble in the 1980 zone consisted mainly of brick, concrete, asphalt, ceramics, and rebar. The material found in this zone tells a different story from the 1964 zone.

The bricks in the 1980 zone boast a variety of markings which can be traced to either companies or specific geographic locations. The markings include: Cooksville, Cooksville Laprarie Brick, Domtar, J.Price, Toronto Brick, Don, T.B. Co., Toronto Brick Co. Ltd., Milton, Canada Brick-Ottawa, Hanson, Canada Brick, Phippen, and Beld. With the exceptions of Canada Brick-Ottawa, Phippen, and Beld, all of the brick is locally sourced.

Another notable feature of the 1980 zone is the active ecological restoration efforts in one of the aquatic waste holding cells, which is adjacent to the Endikement. In this zone, there are obvious efforts to create marsh habitat for birds, fish, and other wildlife (MTRCA 1992, p. 92). The difference between the levels of vegetation present in the 1980 zone and the 1964 zone highlights the distinct stages of ecological regeneration and rehabilitation in these two locations. In this area, we find intentional and strategic

Figure 7.   The worn brick beach in the 1980 zone.

Figure 8.   Concrete slabs arranged to create harbour-like conditions in the 1980 zone.

production of nature, through implementation of the *Tommy Thompson Park Terrestrial Natural Area Enhancement Master Plan* (TRCA 2006), a plan to create and groom habitat for species of concern.

Combined with archaeological fieldwork and analysis of policy and planning records, the fieldwork conducted for this research reveals historical transitions in the city. Inferences from the rubble of the Leslie Street Spit establish a link between the rubble found in the 1964 and 1980 zones and the activities of the City of Toronto during these two time periods. Illustrating the relationship between the rubble (what was destroyed) and the building trends (what was created) during these two years gives insight on the planning preoccupations and ambitions of the city. The stories behind the rubble of the Leslie Street Spit demonstrate that the landscape of the Spit is much more than a nondescript mass of construction waste, but rather a landscape full of memory about the removal of undesirable built resources of the city and their replacement with new forms of architecture.

Brick in the City of Toronto

The discovery of Ontario's aggregate resources through transportation and urban development had a profound influence on the built environment of Toronto, where easily accessible surface clays belonging to the Don and Scarborough beds provided the building materials for much of early construction in the city. Brick became the main construction material used in Toronto after wood buildings were outlawed in Toronto due to repeated fires and the Great Fire of 1904 (Yundt and Augaitis 1992, p. 2). Following the fire prevention by-laws, the widespread use of red and yellow brick as a building material became one of Toronto's most distinct architectural characteristics (Relph 1990, p. 33). The predominance of red and yellow brick in the city is reflected at the Leslie Street Spit where these materials comprise a large portion of the rubble that forms the landscape substrate. Despite its importance to the history and architectural distinctiveness of Toronto, many significant brick buildings have been destroyed in order to clear the path for development and create space for new forms of architecture that reflect the progress of the city.

Many of Toronto's old buildings and streetscapes were sacrificed in an effort to become a modern city. Dendy and Kilbourn write,

> In their enthusiasm for progress the rich and powerful of Toronto, the private and public patrons who have shaped this place over the past two centuries have let go many of our best buildings and streetscapes. Time and time again a lust for the grandest, the latest, or the most fashionable has led to casual neglect or cavalier destruction of our urban heritage. (1986, p. viii)

This sentiment is shared by Eric Arthur, an architectural historian writing during the time of the construction of the new city hall, who explains that "In the march of progress, we have ruthlessly destroyed almost all our older architecture; street names cherished for a hundred years or more have been altered to suit the whims of people on the street..." (Arthur 1964, p. xv). The dramatic alteration of the built environment of Toronto is especially evident in archival photographs depicting a predominance of nineteenth-century brick structures in the downtown core of the city, few of which remain today (Figures 9 and 10).

In terms of the rubble at the Leslie Street Spit, both 1964 and 1980 stand out as years with high levels of destruction, based on the MTRCA (1992) report. What was lost? What activities in the city warranted such high levels of destruction during these two years? The 1960s were a time of modernist planning that largely began with the establishment of the Municipality of Metropolitan Toronto (Metro) in 1953 (Kipfer and Keil 2000, p. 29). In its early stages during the 1960s, the Metro regional government embraced modernist planning principles and used an expertise-based strategy in planning Toronto. As pointed out by

Figure 9.   Looking north from 84 Yonge Street – photograph taken in 1897.
Source: City Engineer's Department, 1897 (City of Toronto Archives, Fonds 200, Series 376, File 1, Item 93).

Figure 10.   Looking north from 84 Yonge Street – photograph taken in 2013.

Kipfer and Keil (2000), the modernist planning approach of this time endangered some of the historically significant areas in the inner city with slum clearance and urban renewal, meaning that many inner-city areas were demolished so that new and modern developments could be put in place (Lemon 1985, p. 151).

One particular example of slum clearance in the 1960s was the Alexandra Park scheme, which targeted an area near Dundas Street and Spadina Avenue for urban renewal. In brief, the Alexandra Park scheme demolished 200 houses on about 9 acres of land and replaced them with 380 units in new style building forms (Sewell 1993, p. 151). The redevelopment took place between 1964 and 1968, focusing on an area that is now bordered by Dundas Street on the north, Cameron Street on the east, Queen Street on the south, and Augusta Avenue on the west. The Alexandra Park scheme was carried out by demolishing the numerous wood and brick houses in the area and reworking the street network; particularly Vanauley Street. A series of photographs were taken in 1965 and 1966 to document the houses that were to be demolished in advance of the new Alexandra Park development. The photographs demonstrate that the row houses were constructed of wood and brick on narrow lots and suggest that they were in general disrepair (Figures 11 and 12), hall-marks of slum neighbourhoods at the time in Toronto.

Despite protests from residents of the area to save the neighbourhood, the proposal went ahead in 1964 and the demolition of the area began soon after (Sewell 1993, p. 154). It was not until the Expropriation Act of 1963 was changed in 1968 that the five-year stretch of urban renewal projects slowed (Lemon 1985, p. 152). Following the demolition of the nine-teenth-century brick and wood housing in Alexandra Park, the area was redeveloped with the majority of the Vanauley Street and Cameron Street being reinvented as what is now the Atkinson Housing Co-op, which was constructed in 1968 (Atkinson Housing Co-operative 2012) (Figures 13 and 14). Alexandra Park is a 410 unit housing project that includes 140 apartments in two medium apartment buildings and 270 townhouses (Atkinson Housing Co-operative 2012).

Figure 11.   Rear view of 15 Vanauley Street – photograph taken in 1966.
Source: Metropolitan Toronto Planning Department, 1966 (City of Toronto Archives, Fonds 200, Series 65, File 672, Item 2).

Figure 12.   Rear view of 62–60 Cameron Street – photograph taken in 1965.
Source: Metropolitan Toronto Planning Department, 1965 (City of Toronto Archives, Fonds 220, Series 65, File 662, Item 9).

Figure 13.   View of 85–89 Vanauley Street – photograph taken in 1939.
Source: Department of Public Works, 1939 (City of Toronto Archives, Fonds 200, Series 372, Sub-Series 33, Item 605).

Figure 14.   Typical streetscape of the Atkinson Housing Co-op – photograph taken from the north-east corner of Vanauley Walk and Felician Place.

The Alexandra Park community is physically enclosed and presents a complete break from the bustle of adjacent neighbourhoods, which include Kensington Market, Chinatown, and Queen West. The plans for the neighbourhood were made with the intention of group-ing people with low income together and enclosing the community from traffic to create an environment that was focused on residents and children of the community (Pravosoudov 2010). The concentration of low income housing was accompanied by an influx of new immigrants to the area as well as a rise in racial tensions and increased violence and drug use. The Alexandra Park development became synonymous with an African–Cana-dian gang called the Project Originals (Pravosoudov 2010). The housing co-operative was formed in 1998 in an effort to address the crime and safety issues of the community. There is currently a development proposal for the revitalisation and redevelopment of Alex-andra Park and the housing co-op. Once again, the master plan proposes to demolish and replace existing homes with a view to enhancing the social fabric and function of this neigh-bourhood (City of Toronto 2012).

While the rubble of the 1964 zone cannot be solely attributed to the Alexandra Park development, given the chronological symmetry between the clearance of this substantial residential neighbourhood and the massive dumping at the Spit, it stands to reason that artifacts from Alexandra Park constitute at least some of the "clean fill" at the Leslie Street Spit. These artifacts are evidence of household displacement and reflect the moder-nist planning principles that were used in the city during the 1960s. Current redevelopment plans for the Alexandra Park neighbourhood call for another round of demolitions, reminis-cent of the initial demolition of the neighbourhood. In contrast, however, the current rede-velopment of Alexandra Park is expected to take between 10 and 15 years to complete and has a particular focus on keeping the community intact through the process. This stands in stark contrast with the 1960s slum clearance initiatives, which took place in four years and resulted in the displacement of its residents and the near complete demolition of the built form of the neighbourhood.

The 1980s represent a very different period for planning in Toronto since the modernist principles that defined the 1960s no longer had strong support in the city (Sewell 1993, p. 174). The year 1980 falls into a boom period for the city where the downtown core experienced a great deal of development – particularly in the form of office towers (Filion 1999, p. 432). Between 1970 and 1980 four downtown landmarks were erected (the CN Tower, the Skydome, the Eaton Centre, and the Harbourfront Centre) and the office surface of the downtown increased by 78% (Filion 2000, p. 173). One block that experienced a great deal of development was the area bordered by Adelaide Street on the north, Yonge Street on the east, King Street on the south, and Bay Street on the west. This block is the site of the Scotia Plaza, a 68-storey office tower with a 14-storey atrium which was constructed in the 1980s and resulted in the demolition of a number of nineteenth-century brick buildings, the materials of which are consistent with the materials at the Spit.

While some brick facades of the nineteenth-century structures were kept, most of the late nineteenth-century and early twentieth-century landmark buildings in this block have been demolished. Three buildings that have survived, at least in facade form, are the National Club, 100 Yonge Street (or the Fairweather building), and 104 Yonge Street. While the facades of these buildings remain, the interior block has been demolished and replaced by the Scotia Plaza and part of the PATH system, an underground pedestrian network that connects much of the downtown core. The Fairweather building and 104 Yonge Street are shown in Figures 15 and 16, which demonstrate how these historic build-ings were incorporated into the Scotia Plaza development.

The demolition and construction of the Scotia Plaza are a reflection of Toronto's 1980s pro-growth strategies that resulted in a construction boom where high-rise office towers

Figure 15.   View of 100 and 104 Yonge Street during construction of the Scotia Plaza – photograph taken between 1972 and 1991.
Source: Ellis Wiley Fonds, 1972–1991 (City of Toronto Archives, Fonds 124, File 3, Item 161).

began to dominate the downtown core. Art Eggleton, Toronto's "business-friendly" mayor was elected in 1980 and during this time planning took an entrepreneurial stance wherein planners were preoccupied with "making deals" with developers and extracting density

Figure 16.   View of 100 and 104 Yonge Street after being integrated into the Scotia Plaza develop-ment – photograph taken in 2013.

bonuses from the downtown office boom (Kipfer and Keil 2002, p. 239). This is also the period where Toronto's economic base became less diversified and more dependent on financial services (Lemon 1985, p. 186). During this time, the planning process made more room for public participation and environmental concerns were increasingly incorporated into plans (Filion 1999, p. 432, 2000, p. 174). While increased public participation curtailed the redevelopment of inner-city neighbourhoods, such as Alexandra Park, construction in the downtown core continued to boom despite planning attempts to calm the growth (Filion 2000, p. 173). This resulted in the destruction of many nineteenth-century brick buildings (such as small theatres, arcades, and bank branches) and their replacement with office towers servicing international finance and trade.

The rubble of the 1980 zone at the Leslie Street Spit represents a shift in the city's planning practice where capitalism and entrepreneurship played more prominent roles. The organisation of the material types and the shaping of the landscape in the 1980 zone of the Leslie Street Spit show that there was considerable planning for the afterlife of the rubble. In this way, the rubble of the 1980 zone reflects the heightened citizen participation and environmental concerns that emerged during this period of urban planning in Toronto (Filion 1999, p. 432). Yet, the materials here also coincide with a shift from small-scale, locally oriented structures towards buildings that positioned Toronto in the global business world.

Creative destruction of Toronto's built environment

Harvey (2006) explains that for a city to become modern it must symbolically create a break with the past in order to discard its old reputation and adopt new ideals. Creating a clear break with the past in order to overcome the failures of the old regime is a tactic that has been used throughout the history of cities (Zukin 2006, p. 113). Often, this break is symbolically represented through architecture, as noted by Kapelos, who writes that "To break with the past and be truly modern demanded a new aesthetic, a new architecture that would ultimately create a humane environment" (2002, p. 42). What is often forgotten in the public imagination is the communities that were displaced during this process. In Toronto, the construction of New City Hall marks an event of creative destruction where new architecture and building techniques were used to introduce modern attitudes in the planning and governance of the city, where the reputation of "Old Toronto" was swept aside in order to allow the city to become modern. Prior to the construction of the New City Hall, the area bounded by Bay Street on the west, Queen Street on the south, University Avenue on the east, and Dundas Street to the north was composed of residential and commercial buildings that dated to the nineteenth century (Goads 1880) (Figure 17). The nineteenth-century brick and wood structures in this area were demolished in advance of the construction of the New City Hall. This reinvented the area and gave a sense of renewal to the entire city. This was effective since old Toronto had the reputation of being cold, quiet, conservative, insular, and "a good place to mind your own business" (Fulford 1995, p. 1).

Many attribute Toronto's shift to a "city that works" to a single event on Monday, 13 September 1965: opening day at the New City Hall and Nathan Philips Square, both designed by the Finnish architect Viljo Revell (Relph 1990, p. 31, Sewell 1993, p. 119, Gandy 2002, Hume 2007, p. 70). As noted by the Bureau of Architecture and Urbanism, "The building of the City Hall transformed the image of the city, and with it, the public perception of Modern architecture" (2002, p. 78). Sewell echoes this sentiment. He writes, "The New City Hall was controversial, but almost immediately it was publicly lauded as the city's successful leap into the future" (1993, p. 119). The building of the New City Hall was a highly symbolic step in the modernisation of Toronto and it marks a departure

City of Toronto Archives, Fonds 1244, fi244_it10083

Figure 17.   Looking north from the intersection of Queen Street and Bay Street in 1910.
Source: William James Family Fonds, 1910 (City of Toronto Archives, Fonds 1244, Item 10083).

from traditional building techniques and materials (Figure 18). Even Revell's choice of concrete as a construction material for City Hall stands in sharp contrast with Toronto's three older City Halls, which were made of brick and locally soured clay (City of Toronto 2010). The construction of the New City Hall symbolised Toronto's leap into the modern world and willingness to discard undesirable elements of its past in favour of structures that suggest readiness for growth.

The Leslie Street Spit is the cumulative result of what was lost during the creative destruction of the city. It is the elements of the city that were discarded so that Toronto could become modern, current, and competitive. The social history embedded in this landscape is obscured by the mythology that the Spit is composed of "clean fill" and the impressive ecological profile of the landscape. The material culture of the 1964 zone is particularly haunting since it is composed of compacted garbage that includes household debris and personal items. The worn brick beach of the 1980 zone and the corroded, artifact laden aggregate from 1964 zone tell the story of Toronto's progress and evolution through modern planning practice and into a post-modern planning framework, including a largely forgotten community that resisted displacement in the wake of this transition.

The Leslie Street Spit: a landscape of memory

The Leslie Street Spit is the ruins of the City of Toronto since it is founded on the rubble that is a product of the ongoing process of creative destruction. Memory and nostalgia for what is lost is a common feature of ruins (Edensor 2005, Huyssen 2006) and the Spit often incites these emotions in its visitors. One way this is evident is through the construction of pathways, sculptures, and other structures out of the rubble by visitors of the Spit (Figure 19). Writing about affordances of industrial ruins, Edensor *et al.* (2012) note that ruins "invite expressive

City of Toronto Archives, Fonds 124   f0124_fl0001_id0147

Figure 18.   New City Hall prior to opening.
Source: Ellis Wiley Fonds, 1945–1966 (City of Toronto Archives, Fonds 124, File 1, Item 147).

physical investigation through the material forms that pre-exist ruination and those that belong to it" (p. 67). It is as if the individuals who created these pathways and structures at the Spit are trying to put the pieces back together to make sense of the rubble. Jones and Evans discuss the importance of materials in creating sites of memory. They write,

> Destroying material traces of sites with deep place associations resets the clock on the embodied relationship between the individual and that environment. Those material sites can act as

Figure 19.   Pathway constructed out of rubble at the Leslie Street Spit.

prompts to recover the memories which have helped to shape those affective connections. (Jones and Evans 2012, p. 2326)

In effect, the creative forms found at the Spit are a clear reminder of the origins of the rubble and suggest ways to imagine the discarded pieces of brick and concrete in their former place in the city. These sculptures help connect the Leslie Street Spit back to the built form of Toronto and serve to remind us of what has been lost.

The rubble of the Leslie Street Spit also illustrates the cycles of decay and renewal, which are strongly associated with ruins. This is noted by Huyssen who explains that, "decay, erosion, and a return to nature" are defining features of ruins (2006, p. 10). The cycle of decay and renewal is a major component of the Spit where the deposition of construction materials and natural regeneration of the landscape are considered a key feature of the "accidental success" of the landscape. The reclamation of disturbed landscapes by nature is one of the most powerful and evocative themes associated with the romaticisation of ruins and the power of nature's ability to renew often stirs feelings of melancholy, hope, and wonder among visitors. The rich ecology at the Spit makes for classic ruins where nature has claimed a derelict landscape and transformed it into a novel ecosystem.

The Leslie Street Spit also functions as a site of conscience and memory. By considering the connections that the Spit has to episodes of destruction in the city we can better understand who and what has been displaced in the recent past. In the case of Alexandra Park, recognising the long-term social issues that resulted from the fast-paced redevelopment of the neighbourhood helps us to appreciate the importance of community consultation and engagement at the outset of future development proposals. This is especially important since the Alexandra Park is undergoing another round of redevelopment in an effort to address the social and structural issues in the community that resulted from the 1960s slum clearance initiatives. Furthermore, by identifying the buildings that were demolished in the downtown core during the 1980s we can evaluate the losses sustained to Toronto's built heritage. This is crucial given the high number of development projects currently ongoing in the downtown core. Many of the nineteenth- and early twentieth-century structures that have been demolished are irreplaceable and represent a permanent loss to Toronto's overall cultural heritage. As noted by the International Coalition of Sites of Memory, "We have seen how – in every part of the world – the past has lessons for our future" (2012). In Toronto, the Leslie Street Spit teaches us that unrestrained development has lasting repercussions that cannot be easily mitigated.

Conclusions

Embedded in the landscape of the Leslie Street Spit is a history of urban socio-economic dispossession. Employing archaeological research methods reveals that the rubble of the Leslie Street Spit is historically associated with the modernist ideals of urban planning prevalent in the 1960s and the private sector driven planning practice of the 1980s. Linking sections of the Spit to contemporary development activities, such as the Alexandra Park scheme or the Scotia Plaza, gives a sense of what was destroyed and then buried at the Spit.

This research assigns a broad context to two sections of the Leslie Street Spit. While this approach yields information about the general social history of the rubble, it has not traced individual artifacts back to specific locations in the city. Future research could include the collection and analysis of artifacts found at the Spit to uncover more information about the neighbourhoods that have been displaced and demolished since 1959.

The Leslie Street Spit's vibrant and celebrated ecology play an important role in masking the underlying history of the landscape. Recognising the Leslie Street Spit as the ruins of Toronto in an unromaticised manner allows one to assign greater meaning to the landscape. Viewing the Spit in this light allows it to function as a site of memory in addition to being a site of urban wilderness. As noted by Herrington, "Since landscapes can contain tangible artifacts of the past, they have played a vital role in stimulating memories and associations" (2009, p. 72). At the Leslie Street Spit, memories and associations with the city are expressed as people create structures out of the rubble. This activity is most pronounced at the southern edge of the 1964 zone where a series of pathways and sculptures have been erected that uses the discarded brick, concrete, and rebar. These structures recall the former use of the rubble and make the social history of the landscape undeniably clear despite the urban development processes that make it easy to forget who and what existed in the recent past.

Acknowledgements

The authors would like to thank various people for their contributions to this project. Thanks are due to Anatolijs Venovcevs and Blake Williams for their valuable support. We would also like to express our great appreciation to our reviewers for their constructive and insightful recommendations.

References

Arthur, E., 1964. *Toronto: no mean city*. Toronto: University of Toronto Press.
Atkinson Housing Co-operative, 2012. *Our history, Alexandra Park to atkinson co-op* [online]. Available from: http://www.atkinsonhousingcoop.com/history.htm [Accessed 10 January 2013].
City Engineer's Department, 1897. *Looking north from #84 Yonge Street*. City of Toronto Archives, Fonds 200, Series 376, File 1, Item 93.

City of Toronto, 2010. *Toronto city hall: a brief History* [online]. Available from: http://www.toronto. ca/city_hall_tour/history.htm [Accessed 17 September 2010].

City of Toronto, 2012. *Alexandra park official plan amendment & rezoning* [online]. Available from: http://www.toronto.ca/planning/alexpark.htm [Accessed 15 January 2013].

Dendy, W. and Kilbourn, W., 1986. *Toronto observed: its architecture, patrons, and history.* Toronto: Oxford University Press.

Department of Public Works, 1939. *85–89 Vanauley Street: see 527.* City of Toronto Archives, Fonds 200, Series 372, Subseries 33, Item 605.

Edensor, T., 2005. The ghosts of industrial ruins: ordering and disordering memory in excessive space. *Environment and Planning D: Society and Space*, 23 (6), 829–849.

Edensor, T., *et al.*, 2012. Playing in industrial ruins: interrogating teleological understandings of play in spaces of material alterity and low surveillance. *In*: A. Jorgensen and R. Keenan, eds. *Urban wildscapes.* London: Routledge, 65–79.

Ellis Wiley Fonds, 1945–1966. *Buildings, part 1.* City of Toronto Archives, Fonds 124, File 1, Item 147.

Ellis Wiley Fonds, 1972–1991. *Buildings, part 3.* City of Toronto Archives, Fonds 124, File 3, Item 161.

Filion, P., 1999. *Rapture of continuity? Modern and postmodern planning in Toronto.* Oxford: Joint Editors and Blackwell Publishers Ltd.

Filion, P., 2000. Balancing concentration and dispersion? Public policy and urban structure in Toronto. *Environment and Planning C: Government and Policy*, 18 (2), 163–189.

Foster, J., 2007. Toronto's Leslie Street Spit: aesthetics and the ecology of marginal land. *Environmental Philosophy*, 4 (1&2), 117–133.

Foster, J. and Sandberg, L.A., 2004. Friends or foe? Invasive species and green space in Toronto. *Geographical Review*, 94 (2), People, Place, & Invasive Species, 178–198.

Fulford, R., 1995. *Accidental city: the transformation of Toronto.* Toronto: McFarlane Walter and Ross.

Gandy, M., 2002. *Concrete and clay: reworking nature in New York city.* Cambridge: The MIT Press.

Goads, 1880. *Goads fire insurance plans, plates 27, 28, 35, and 36. City of Toronto Archives* [online]. Available from: http://www.toronto.ca/archives/maps/goads-1880-index.htm [Accessed 13 February 2013].

Harvey, D., 2006. *Paris, capital of modernity.* New York: Routledge.

Herrington, S., 2009. *On landscapes.* New York: Routledge.

Hume, C., 2007. New city hall. *In*: M. McLelland and G. Stewart, eds. *Concrete Toronto: a guide to concrete architecture from the fifties to the seventies.* Toronto: Coach House Books, 70–77.

Huyssen, A., 2006. Nostalgia for ruins. *Grey Room*, 23, Spring, 6–21.

International Coalition of Sites of Conscience, 2012. *Approach, international coalition of sites of conscience* [online]. Available from: http://www.sitesofconscience.org/approach/ [Accessed 5 August 2013].

Jones, P. and Evans, J., 2012. Rescue geography: place making, affect and regeneration. *Urban Studies*, 49, 2315–2330.

Kapelos, G., 2002. A modern vision for Toronto. *In*: Bureau of Architecture and Urbanism, ed. *Toronto modern: architecture 1945–1965: catalogue of the exhibition with critical essays.* Toronto: Coach House Books, 38–45.

Kipfer, S. and Keil, R., 2000. Still planning to be different? Toronto at the turn of the millenium. *disP – The Planning Review*, 36 (140) 28–36.

Kipfer, S. and Keil, R., 2002. Toronto Inc? Planning the competitive city in the new Toronto. *Antipode: A Journal of Geography*, 34 (2), 227–263.

Lemon, J., 1985. *Toronto since 1918: an illustrated history.* Toronto: James Lorimer & Company, Publishers.

Metropolitan Toronto and Region Conservation Authority (MTRCA), 1992. *Tommy Thompson Park master plan & environmental assessment* [online]. Available from: http://www.trca.on.ca/enjoy/ locations/tommy-thompson-park.dot [Accessed 3 May 2010].

Metropolitan Toronto Planning Department, 1965. *Alexandra Park: Cameron Street.* City of Toronto Archives, Fonds 220, Series 65, File 662, Item 9.

Metropolitan Toronto Planning Department, 1966. *Alexandra Park: Vanauley Street.* City of Toronto Archives, Fonds 220, Series 65, File 672.

Ontario Ministry of the Environment (MOE), 1982. *Lakefill quality study: Leslie Street Spit, city of Toronto*. Toronto: Environment Canada.

Open Street Map, 2013. *Leslie Street Spit 45°63'06" N 79°33'94"* [online]. Available from: http://www.openstreetmap.org/#map=14/43.6307/-79.3394 [Accessed 9 August 2013].

Pravosoudov, M., 2010. *Cities for people – Alexandra Park* [online]. Spacing Toronto. Available from: http://spacing.ca/toronto/2010/06/22/cities-for-people-alexandra-park/ [Accessed 15 January 2013].

Rathje, W. and Murpy, C., 2001. *Rubbish! The archaeology of garbage*. Tuscon: The University of Arizona Press.

Relph, E., 1990. *The Toronto guide: the city, metro, the region*. Prepared for the Annual Conference of the Association of American Geographers, Toronto.

Renfrew, C. and Bahn, P., 2004. *Archaeology: theories, methods, and practice*. New York: Thames & Hudson Ltd.

Sewell, J., 1993. *The shape of the city: Toronto struggles with modern planning*. Toronto: University of Toronto Press.

Toronto Port Authority, 2012. *Notice to contractors and truckers: Leslie street lakefill site* [online]. Available from: http://www.torontoport.com/TorontoPortAuthority/media/TPASiteAssets/PDFs/Lakefill-Regulation-2012-June-6.pdf [Accessed 28 July 2013].

Toronto Region and Conservation Authority (TRCA), 2006. *Tommy Thompson Park terrestrial natural area enhancement master plan*. Toronto: Toronto Region and Conservation Authority.

Toronto Region and Conservation Authority (TRCA), 2010. *Tommy Thompson Park: Toronto's urban wilderness* [online]. Available from: http://www.trca.on.ca/enjoy/locations/tommy-thompson-park.dot [Accessed 5 November 2010].

William James Family Fonds, 1910. *Looking north from Queen and Bay streets*. City of Toronto Archives, Fonds 1244, Item 10083.

Yokohari, M. and Amati, M., 2005. Nature in the city, city in the nature: case studies of the restoration of urban nature in Tokyo, Japan and Toronto, Canada. *Landscape and Ecological Engineering*, 1, 53–59.

Yundt, S.E. and Augaitis, D.B., 1992. *From pits to playgrounds: aggregate extraction and pit rehabilitation in Toronto – A historical review*. Prepared for the Ontario Ministry of Natural Resources. Toronto: Queen's Printer for Ontario.

Zukin, S., 1991. *Landscapes of power: from detroit to disney world*. Berkley: University of California Press.

Zukin, S., 2006. David Harvey on cities. *In*: N. Castree and D. Gregory, eds. *David Harvey: a critical reader*. Oxford: Blackwell Publishing, 102–120.

Re-presenting transgressive ecologies: post-industrial sites as contested terrains

Joern Langhorst

College of Architecture and Planning, University of Colorado Denver, Denver, CO, USA

In an age where urban agglomerations are shaped by rapid economic and demographic changes, post-industrial sites are interpreted as symbols of the failure of the industrial age and location of essential, sustainable and environmentally just renewal. Both understandings are based on constructions of "culture" and "nature" as dichotomic and normative. This paper proposes an alternative framework that constructs post-industrial sites as locations for the ongoing negotiation of human and non-human processes, as passive ground and active agent in the constant and continuous remaking of place. Landschaftspark Duisburg-Nord and the High Line in New York City are highly visible examples that engage and showcase successional ecologies in fundamentally different ways, but both ultimately posit "ecology" as the redemptive agent for emergent *post*-postindustrial places. While Duisburg-Nord engages emergent ecologies as an exploratory, open-ended and fluid interaction between human intervention and non-human process, the High Line replaces actual successional ecologies with artificial ones based on picturesque traditions. These representations and aesthetications of ecological processes construe both sites as icon and instrument of deterministic-hegemonial and transgressive-marginal agendas, emphasising their role in postmodern and post-colonial discourses on the nature of urbanity and sustainable urban renewal – discourses central to current ideas of "landscape urbanism" and "ecological urbanism". The "right to the city", in contemporary parlance, implicitly comprises a "right to urban nature". "Aesthetics of (ecological) performance" vs. "(ecological) performance of aesthetics" will serve as a key dialectic in the analysis of the relationship between environmental aesthetics and environmental justice as it is played out on post-industrial sites and attempts to construe versions and visions of "urban nature".

Introduction and methodology

This paper investigates two of the most iconic and discussed design projects on post-industrial sites: Landschaftspark Duisburg-Nord (Duisburg, Germany, 1989–1999), and the High Line (New York City, 2007–ongoing).

It interrogates different constructions of "urban nature" and emergent ecologies on iconic and highly visible post-industrial sites in the context of past and current discourses on social and environmental justice, urban renewal and sustainability. It argues that the main relevance of emergent ecologies on such sites lies not in their physical-ecological functions and per-formances, but in their *aesthetic* and *representational* agency. While Landschaftspark Duis-burg-Nord and the High Line are 20 years apart and different in context, scale and complexity, both posit "ecology" as a central redemptive agent for post-industrial sites and urban renewal. They differ vastly in their treatment, aesthetication and instrumentalisation of emergent ecol-ogies, and the resulting constructions of "urban nature" reveal distinctively different ideol-ogies and agendas that can be considered prototypical.

The paper first locates design projects on post-industrial sites in the larger contexts and discourses on social and environmental justice, urban redevelopment and sustainability, "nature" and "culture", and identifies "urban nature", expressed through emergent ecologies, as a central concept. It posits their main agency as *aesthetic* and *representational*, and intro-duces the "aesthetics of ecological performance" vs. the "performance of aesthetics" as a key dialectic for the analysis of both projects. The traditions of the "picturesque" and "sublime", concepts that were central to the field and discipline of eighteenth- and nineteenth-century landscape architecture, are discussed as continuing influences in the design of post-industrial sites, despite assertions to the contrary by the designers of Duisburg-Nord and the High Line and the current discourses on landscape urbanism and ecological urbanism.

The subsequent analyses and interpretations of Landschaftspark Duisburg-Nord and the High Line are based on extensive literature reviews, the author's observations and experi-ences from multiple site visits, and informal interviews with visitors, public officials, some of the designers and residents of surrounding neighbourhoods.[1]

These interpretations greatly expand past and current critiques that habitually focus on ecological-physical landscape benefits and performances, and apply established theoretical concepts: (the "right to the city" and attendant concepts of social, environmental and aesthetic justice; representational and aesthetic agency of landscape; heterotopian spaces; the aesthe-tication and instrumentalisation of emergent ecologies; as well as a critical analysis of pictur-esque ideology and aesthetics) to both sites for the first time, synthesising new theoretical frameworks to critically evaluate approaches to projects on post-industrial sites and in con-tested urban conditions. Additionally, these interpretations foreground the aesthetic agency of post-industrial sites in general and of emergent and transgressive ecologies in particular, and extend the concept of the "right to the city" to a *right to urban nature*. The understanding of post-industrial sites as *simultaneous ruins of nature and culture* serves as a key framework to uncover and decode the involvement and co-authorship of human and non-human pro-cesses beyond ideological constructs of "nature" and "culture" and thus construe alternate, inclusive and empowering versions and visions of "urban nature".

Landschaftspark Duisburg-Nord and the High Line are found to be exemplary and representative of particular attitudes towards and constructions of human–nature relation-ships and processes in general, and urban nature in particular, with distinct consequences for the agency, efficacy and potential of post-industrial sites in general.

The conclusion investigates these underlying ideas about "urban nature" further and identifies three major concerns that should guide the discourses and practices relating to the increasing role of post-industrial sites in general and emergent ecologies in particular in the re-imagining and re-making of contemporary cities.

Background

Post-industrial sites are simultaneously iconic reminders of the ongoing de-industrialis-ation, and locations to re-imagine, reinvent and recover landscapes as agents for cultural, social, economic and ecological change. Frequently seen as aesthetically sublime,[2] they are interpreted as symbols of the failure of the industrial age and location of essential, sustainable and environmentally just urban transformation. Both understandings mythologise and value such places based on a fundamental construction of "culture" and "nature" as dichotomic and normative.[3]

This paper challenges this dichotomy by exploring post-industrial sites as an ongoing negotiation and interaction of human and non-human processes and positing them as *simultaneous ruins of culture and nature*. Emergent ecologies are a visible expression of this interaction, and their treatment plays a key role in understanding and addressing problems and potentials of post-industrial urbanisations.

Post-industrial sites as contested terrains

The most visible discourses on post-industrial sites emphasise issues of environmental justice, of pollution mitigation, of adaptive reuse, and, in the context of urban redevelopment and renewal, of benefits and performances based on ecological functions and processes. These discourses propose concepts such as "landscape urbanism", "sustainable urbanism" and "ecological urbanism"[4] as conceptual and practical alternatives to the mainstream forces of urban development and suggest, in different ways, ecology as redemptive agent. Landscape urbanism, and its latest iteration, ecological urbanism, in particular emphasise process over finished static form, and posit landscape and place as both results of cultural production and agents for cultural change. Consequentially, their approaches towards intentional environmental change attempt to engage the inherent open-endedness and dynamic qualities of human and non-human processes and systems. While these approaches include many interests, processes, forces and factors excluded in more traditional planning approaches, landscape urbanism, in particular, has been criticised for fetishising process and possibility over tangible outcomes, and being much more descriptive of existing conditions than generating ideas about future ones (Talen 2010).

Traditional and new hegemonial agendas of global development and capitalisation, with their attendant less obvious and less public discourses, continue to focus on political and economic interests involved in the complex processes of urban renewal, gentrification, redevelopment and rebranding, producing economically, socially and environmentally uneven and unjust conditions (Harvey 1973, 1996, Smith 1984). In this context, emergent ecologies and "urban nature" are frequently instrumentalised and play multiple important roles.

These competing discourses construct post-industrial sites as highly contested terrains, and force a reconsideration of ecology from an objective construct of descriptive science focusing on biophysical aspects to a contextually complex hybrid, involving cultural, political, economic and biophysical aspects. Post-industrial sites themselves could then be decoded as physical manifestations of such discourses.

Such sites have the potential to be at the centre of conversations about urbanity and urban transformation, and the processes, roles and treatment of emergent ecologies can be posited as a key issue. Consequently, spatial design and planning projects that address post-industrial sites have attained a high degree of visibility, in particular if they employ a significant public open-space or green-space component. While criticism has mostly focused on either aesthetic appearance or ecological performances, their location and social, cultural and economic contexts suggest the need to emphasise issues of environmental justice.

Urban nature(s) and environmental justice

This paper argues that the "right to the city" (Harvey 1973, Lefebvre 1996, Mitchell 2003), historically defined as an issue of social and (to a degree) environmental justice,[5] also implicitly comprises a *right to urban nature*.

The term "urban nature" is in itself ambiguous, but frequently used to describe spontaneous vegetations and other expressions of non-human processes in densely built and controlled urban developments. As urban nature is continuously produced by human and non-human processes and agents in varying spatial, material and temporal contexts, it would be more appropriate to use the plural "urban natures".

Access to and experiences of urban nature or of "nature" in general is one of the primary concerns, as most post-industrial sites occur in neighbourhoods that are characterised by a lack of parks and other open space, below-average environmental quality, and frequently inhabited by communities that are affected by the very processes of industrialisation and deindustrialisation that created such sites in the first place. As urban nature is produced (or is allowed to produce itself) on post-industrial sites, questions of the types of nature, how they are distributed and how they impact the life of people in their proximity become a concern (Foster 2010, p. 318, see also Hester 2006).

Emergent ecologies can be considered the most authentic[6] elements within many concepts of urban nature, as they are the physical expressions of non-human processes that are not controlled by human maintenance regimes. They may be interpreted as nature's "spatial tactics"[7] and acts of resistance, occupying marginal and marginalised spaces that are less or not at all controlled by hegemonial powers of development and capital. Such emergent ecologies, most often manifested in early stage successional vegetation, then could be more appropriately understood and labelled as *transgressive ecologies*.[8]

In this context, the main agency of post-industrial sites might not lie primarily in their actual and potential ecological functions and performances,[9] but more so in their *aesthetic* and *representational* function.

Aesthetics of landscape and landscapes of representation

Aesthetics are defined here as complex human–environment relationships (Berleant 1992), not just a normative visual practice, and proposed as everyday "aesthetic experience", in particular as the ability to interact with urban nature, with non-human processes and their physical expressions. Not only is this experience considered enjoyable and contributes to the overall quality of life, it is also a critical precondition to imagine and articulate alternative relationships between the individual and collective self and urban nature and nature in general (e.g. Tuan 1974, Louv 2005, Hester 2006).

The representational agency (Cosgrove 1998, Corner 1999) of landscape itself, its ability to represent ideas and ideologies, is particularly potent if it is intentionally designed. Its representational capacity relies heavily on aesthetic qualities, often foregrounding passive visual reception (appearance) over all other forms of sensory experience. The way in which underlying ideas inform and correspond to spatial and material form, and determine regimes of maintenance and control, is frequently deeply encoded, and not obvious. The decoding of designed and professionally curated landscapes depends on a particular situated[10] knowledge of aesthetic and ideological traditions of landscape architectural or architectural design, and of the long-standing complicity of spatial design and planning professions with hegemonial agendas. However, designed landscapes are – due to their experiential qualities – capable of representing ideas, ideologies and underlying values independent of an audience's ability to decode them, an ability that grants insidious power. In the context of post-industrial landscapes their ability to perform and represent ideas about the relationship between human and non-human processes, between "nature" and "culture" is of particular interest.

Aesthetics of (ecological) performance vs. *(ecological) performance of aesthetics* will serve as a key dialectic in the paper's analysis of the relationship between environmental aesthetics and environmental justice as it is played out on post-industrial sites and attempts to construe and instrumentalise versions and visions of "urban nature" in the context of urban transformation. "Aesthetics of (ecological) performance" foreground the processes that occur on a site as they produce somewhat open-ended aesthetic experiences, while the (ecological) performance of aesthetics attempts to create and control a particular aesthetic experience and constructs or manipulates ecological processes and the resultant conditions to that end.

This dialectic can be translated into two competing frameworks that are evident in the designs of Landschaftspark Duisburg-Nord (Duisburg, Germany) and the High Line (Manhattan, New York), two of the most iconic and most widely discussed examples of post-industrial sites. These frameworks allow exploration of the constitutive function of the picturesque tradition and its aesthetic and ideological aspects in the construction of "urban nature", in particular as it addresses the appearance, performance and agency of emergent successional and transgressive ecologies.[11]

The picturesque heritage

The concept of the "picturesque", rooted in enlightenment romanticism, was fundamental in the development of landscape architecture as a field, discipline and practice in the eighteenth and nineteenth century, translating "techniques, formal strategies, and subject matter of Romantic paintings into the design of landscapes" (Herrington 2006, p. 23).[12]

While it is used to shape the most recognisable appearances of designed landscapes (Central Park in New York City might be the most visible example), and all but disappeared as a style, its aesthetic operations and ideological agency are still influential, and particularly relevant to the designs of the case studies.[13]

Picturesque landscapes were quite literally "built" after paintings or images depicting idealised versions of nature. Operating somewhere between Edmund Burke's categories of "beautiful" and "sublime" (Meyer 1998, Hunt 2000, Hunt and Willis 1975, 2002, Herrington 2006, p. 24), the resulting landscapes covered the gamut between the "pastoral" – inhabited rural, idyllic landscapes, emphasising the comfortable and relatively tame – and the "sublime" – wild nature, vast and powerful, inspiring terror and awe. The underlying assumptions about the nature and relationships of human and non-human processes are by no means relegated to (or literally build into) works of landscape architecture in the picturesque style of the eighteenth and nineteenth century, but insidiously expressed through and represented by much more contemporary designed landscapes.

The representational agency of designed landscapes in general, and post-industrial sites in particular, relies heavily on the accultured concepts of romantic nature as either pastoral or sublime. In order to incorporate sublime notions into any design, the wild and awe-inspiring qualities are scaled down, translated into different spatial and material expressions, and ultimately "domesticated", thus removing the threat to human comfort and the systems of social and spatial order operating in and around the sites.

Meyer (1998) emphasises the role of "sublime sentiments" in Richard Haag's design for Gasworks Park, an iconic ruin of a gasification plant located across from Downtown

Seattle. A project on a post-industrial site, it has been hugely influential as the first design (1971–1973) to retain the abandoned and derelict industrial structures, and to epitomise a new "post-industrial" rendering of the sublime.

The romantic ruin and simultaneous ruins of culture and nature

Post-industrial sites, in particular those with iconic remnants of buildings, have a long tradition of being conceptualised as ruins and as vestiges of a better, more prosperous past. Ruins and wilderness played a central role in picturesque aesthetics, producing experiences of nature and its effects on human artefacts as the quintessential "sublime".[14] It is rooted in aestheticised concepts of nature as the "other" and as "dehumanised" (Evernden 1992, Turner 1994). The "othering" of nature occurs within two frameworks central to the Enlightenment agendas of political emancipation and self-realisation: rationalism frames nature as "measurable", a pretext to its valuing as a resource to be exploited, romanticism posits nature as the "ideal other" that models human nature. In this context, nature, as the source of awe-inducing "sublime" experiences, needs to be considered as a problematic "source of authority" (Spirn 1997). Consequentially, the picturesque is not an immutable, innocuous, nostalgic and idyllic representation, but an insidious ideology involved in "naturalizing power" and in "deifying human systems of power" (Robinson 1988, Mitchell 1994).

The capacity of ruins to convey notions of the sublime is rooted in making visible the passing of time ("memento mori") and the efficacy of the destructive forces of non-human processes undoing the works of the human hand (for critical positions on the role of ruins in the construction of civilisation and cultural memory see Arendt 1958, Foote 2003, Huyssen 2003, Marot 2003, Nora 1989). The ongoing interaction of human and non-human processes in ruins yields continuously changing conditions and experiences, antithetical to the attempts to create an unchanging and "perfect" landscape. This static understanding of landscape is characteristic of picturesque traditions, as seen in the romantic (and romanticised) classical ruins frozen in the perfect state of decay in eighteenth-century landscape paintings – paintings that were subsequently translated into the designed landscapes of stately homes and large parks of the eighteenth and nineteenth century. While such landscapes attempted to carefully hide the human hand, they were in fact tightly managed, maintained and controlled, designed to provide a particular visual experience that communicated a distinct socio-political order and its attendant values – a *performance of aesthetics*.

This picturesque rendering and understanding of ruins and nature is opposed to the concept of *simultaneous ruins of nature and culture*.

Non-human processes – the acts of weathering, of erosion, of the development of spontaneous vegetation and emergent ecologies ultimately win over human attempts to maintain or restore order (see Figure 1). These more-than-human processes are conventionally considered responsible for the ruination of human edifices and transgressing on the very notion of culture and civilisation.

The ongoing negotiation of human and non-human processes does not just produce cultural ruins – any act of building, of impositions of human ordering systems, conversely produces ecological ruins or ruins of nature. A road paved with concrete might be considered a total ruin of nature (and a complete edifice of culture), until cracks appear, occupied by emergent vegetation. Weathering and vegetation development will further "unbuild" the cultural and "build" the natural until the concrete pavement is completely broken down and assimilated into the soil. Any place, site and landscape then can be described *as simultaneous ruins of culture and nature* (Langhorst 2004, 2012), and

Figure 1. Iconic ruins of the blast furnace complex in Landschaftspark Duisburg-Nord. *Source*: Photograph by the author.

occupies a position on a gradient between complete ruins of nature and complete ruins of culture (or, between complete edifices of culture and complete edifices of nature).

This concept is particularly well suited to analyse and conceptualise post-industrial sites, as it does not privilege either human or non-human process.

Landschaftspark Duisburg-Nord: post-industrial sites as productive heterotopias and as simultaneous ruins of nature and culture

Landschaftspark Duisburg-Nord (Landscape Park Duisburg – North), located at the western edges of the Ruhrgebiet industrial area in Germany, a conurbation of 17 cities and 2.5 million inhabitants, comprises 230 ha of former blast furnaces, turbines, ore bunkers, machine halls, foundries, gas tanks, storage areas, workshops, a sewage treatment plant, a coking plant and a network of rails and roads. When the plant was closed in 1985, 8000 steelworkers were laid off, but a majority continued to live with their families in workers' housing schemes neighbouring the plant. It quickly became the figurehead project[15] of the IBA (Internationale Bauausstellung: International Building Exhibition) Emscherpark, a 10-year, 4 billion DM[16] project for "a regional policy programme for sustainable ecological, economic and aesthetic renewal for an industrial region that had been exploited to the maximum" (Weilacher 2008, pp. 104–105). The IBA employed a distinct planning process: it provided an "open forum", combining a top-down regional planning approach with bottom-up local initiatives, promoting public–private partnerships, and decentralising planning and design decisions, employing concepts of the post-Fordist city (see Schuhmacher and Rogner 2001).

The qualities of the industrial remnants are similar to those found in depictions of ruins in eighteenth- and nineteenth-century landscape paintings, evoking "sublime" experiences.

Between 1985 and 1989 the former factory site lay fallow, and was considered "terra incognita" and "an ecological disaster area which nature was slowly winning back for herself over the years" (Weilacher 2008, p. 105). In 1990, after one year of going through an elaborate planning and design process with considerable citizen and stakeholder involvement, Peter Latz + Partner's design was chosen. Avoiding the obvious focus on land renewal and recycling, and on "preserving iconic industrial monuments,[17] it embraced the IBA's process-driven, functional transformation of complex landscape structures with an inevitably high proportion of unpredictability" (Weilacher 2008, p. 109). The design is exceedingly complex, but is rooted in an inclusive engagement of multiple competing and alternate processes and ordering systems as ongoing forces that affect form, materiality and performance.

First and foremost, the design avoids attitudes and normative judgements based in the othering of nature and non-human processes (Evernden 1992, Soule and Lease 1995), and reconceptualises nature as a pre-existent and inevitable co-author of the site, thus creating (or at least acknowledging) a "state of permanent unsettling" (Hetherington 1997) that forces ongoing critical reflection on the relationships between human and non-human processes.

Landschaftspark Duisburg-Nord preferences neither human acts of civilisation and ordering nor the non-human processes bent on imposing their own order. The presence, entanglement and co-authorship of both human and non-human process is as integral as it is visible, and makes the park a quintessential opportunity to explore and experience post-industrial landscape as *simultaneous ruins of culture and nature* (Langhorst 2004, 2012).

Over time, they create a complex mosaic of different conditions on the gradients of planted–spontaneous and maintained–not maintained, and blur–distinguishing lines between human and non-human systems as sole or even dominant authors.

In Duisburg-Nord existing transgressive/emergent ecologies and spontaneous vegetation were not excluded or replaced, but integrated into the design – as critical and expressions of non-human processes, and amended by planted vegetation, creating a continuous process that blurs any distinguishing lines between nature and culture (see Figure 2).[18] Nature is not just posited as the antagonist of human civilisation, involved in the destruction of human systems of order and production, but a dynamic force to be continuously engaged and incorporated in any human endeavour. The resulting relationship is inevitably dynamic in itself, defying any attempt to represent it or its constituent forces as static. This experience of multiple simultaneous non-human and human processes forces an interpretation of post-industrial sites as neither utopian or dystopian, but heterotopian – as sites of simultaneous competing ordering systems that are in a constant state of becoming (Hetherington 1997).[19]

> ... whereas heterotopia's ad hoc inclusivity and openendedness portends a disturbing and bewildering prospect, it also systematically denies singularity, totality, determinacy, and hierarchy. As a "structured heterogeneity" such a complex field is neither chaotic nor ordered, but free and organic.

Figure 2. Interaction of spontaneous vegetation and transgressive ecologies with planted and maintained systems.
Source: Photographs by the author.

Thus, a truly ecological landscape architecture might be less about the construction of finished and complete works, and more about the "processes", "strategies", "agencies", and "scaffold-ings" – catalytic frameworks that might enable a diversity of relationships to create, emerge, network, interconnect, and differentiate. (Corner 1997, p. 102)[20]

This dynamic state is inherently open-ended, positing challenges for professions and disci-plines of spatial design that traditionally rely on concepts of nature as "homeostatic" and emphasise equilibrium or predictable conditions. The philosophical error of assuming that nature is essentially unchanging has led to actual damage to ecosystems managed by well-meaning believers in natural homeostasis (Botkin in Turner 1994, p. 42).

On its other extreme, the anticipated range of variability of non-human systems can be used to avoid making decisions on intentional change and resort to a position rooted in the normative power of the factual. Berrizbeitia (2001) suggests to engage the inherent uncer-tainty by being "precisely open-ended" rather than "vaguely loose".[21]

Consequently, Landschaftspark Duisburg-Nord is posited by the designer and many critics as an ongoing experiment – the temporal becomes as important as the spatial. It emphasises the interrelations of processes and forms. It favours maintenance regimes that "edit" ongoing processes over the more radical interventions of building and removal. The rendering visible of a significant number of processes, allowing for direct interactions with the processes and their impacts over time are important elements of the design, as is the repurposing and reuse of existing structures within new or different net-works and processes. The sophisticated development of a stormwater management and treatment system repurposing existing industrial structures (see Figure 3), and the use of

Figure 3.   Stormwater infrastructure, Duisburg-Nord.
Source: Photograph by the author.

spontaneous and planted wetland vegetation is one prominent example that also illustrates the design's mandate for "process-driven, functional transformation of complex landscape structures" (Weilacher 2008, p. 109).

Infrastructural functions are "layered" into existing systems by combining the inventive and adaptive reuse of existing structures with careful additions to make the necessary physical and functional connections.

It is absolutely critical to understand that Duisburg-Nord is not about "ecological restoration" – it is about creating "new and interactive alliances" (Corner 1999, p. 4) between human and non-human systems – in other words, new and hybrid socio-ecological assemblages (Kaika 2009, Swyngedouw 2010). These assemblages privilege neither natural nor human systems, and "soften" or alleviate environmental and social inequalities based in the segregated functionalism and efficiency of modernist-capitalist landscapes of production.

Despite its structural complexity and programmatic density, Landschaftspark Duisburg-Nord does not feel over-programmed. Flexible, imaginative uses are tolerated, and where they occur frequently, spatial conditions might be adapted to accommodate them (see Figure 4). Even activities considered socially transgressive are mostly tolerated, unless they interfere with other people's ability to enjoy the place.[22] Very few areas are off limits – access onto the industrial structures is made possible, applying a very commonsensical approach to safety and risk management. Duisburg-Nord strikes an intriguing balance – much different from many other projects on similar sites, it does not render all structures or vegetation "precious". Some parts of the site, in particular structures and buildings that are reused in sanctioned ways, as offices, restaurants, gallery spaces are obviously off limits for activities that involve leaving distinct marks or the slow destruction. Edensor (2005, p. 27) describes the attraction of industrial ruins as allowing a "radical engagement with the material world" made even more intriguing by the clear transgression on social conventions – climbing on structures that are clearly implying this as a forbidden activity, as well as the slow destruction and rearrangement of leftover debris piles are just the more visible part of a wide array of activities that

Figure 4.   Human activities in Duisburg-Nord. Climbing wall in former ore bunkers, performance at the Piazza Metallica.
Source: Photographs by the author.

may appear to be dystopian signs of an anti-authoritarian, to constitute a nihilistic rebuke to conventions of civil order and responsibility, but can alternatively be considered as entirely pleasurable adventures typified by a liminal letting go of the restraints which organize social life. (Edensor 2005, p. 27)

Graffiti are not removed and form a palimpsest that renders legible a wide range of assertions of identity, territoriality and protest. Hidden and hard to access areas of the site support small structures – the hideouts of groups of teenagers, or small vegetable patches, tended to by their unknown gardeners.

Visitors and users become agents in the ongoing change of the site, and are not just relegated to the role of passive consumer in a post-industrial theme park. Their aesthetic experience of human and non-human processes is an immersive, interactive one – and their impact on plants and other materials makes them co-authors of the place they occupy. The manipulation of nature in the picturesque era was relegated to landowners and their empowered agents. In a more contemporary sense, the different iterations of "nature", in particular in dense urban contexts, are quickly considered precious, and off limits to the general population. Tight regulations, as in parks or in all kinds of nature preserves, are discouraging inhabitation and interaction, and emphasise the acts of looking, of beholding an aesthetic object. Areas of "wildness" that can be occupied are usually at a distance from dense, urban neighbourhoods, and frequently out of reach for the communities that do not have the financial means or transportation to regularly access such sites. In this context, Duisburg-Nord fulfils a critically important function for the surrounding neighbourhoods, and, because of its complexity, diversity of conditions and sheer scale, facilitates a wide range of opportunities to encounter an equally wide range of "urban natures" – and, in fact, participate actively in the making of a landscape, place and space (see Tilley 1994, Ingold 2000, Creswell 2004).

The variety of uses, activities and experiences was largely unchoreographed and emerged over time. Many activities were originally considered transgressive, and became officially sanctioned over the years. The overall level of control over visitor behaviour is comparatively low. Basic rules of securing structures have been followed, but the park still invites exploration, transgression, and "niching" for particular groups or individuals.

Participation, possibility, permissiveness and open-endedness characterise Duisburg-Nord's inclusion of the human and non-human component, and create a productive heterotopia that continuously produces new interactions between "nature" and "culture" – interactions that invite and force ongoing critical reflection.

While Duisburg-Nord retains the appearance and experience of the "sublime" so central to picturesque aesthetics, it by far transcends a merely visual reception, instead providing an *aesthetics of human and non-human performance*:

In addition, it openly frames and reveals the ongoing interaction of human and non-human processes, yielding continuously changing conditions and experiences. This approach is antithetical to the attempts to create an unchanging and "perfect", tightly controlled and maintained landscape typical of picturesque aesthetics.

Not only is Duisburg-Nord based in the idea of landscape as constantly changing, but it also posits the human experience of and involvement in this change as central, facilitating and encouraging much more active, immersive and participatory experiences beyond the passive (and voyeuristic) practice of looking.[23] The significant aesthetic appeal of Duisburg-Nord's iconic ruins makes it easy to overlook that it continues to be a working landscape, even beyond the many forms of active and passive recreational activities. The experience of its actual ecological performances and benefits, e.g. as part of a stormwater management system or experiments in phytoremediation are as much part of its aesthetic as the photogenic remnants of its blast furnaces.

The high line: post-industrial sites as picturesque artefacts

The High Line on the West side of Manhattan started as an elevated freight rail track in the 1930s and was eventually decommissioned in 1980. Largely left unmaintained, spontaneous successional ecologies developed, mostly consisting of draught-tolerant species. Iovine (2009, p. D7) describes the High Line during this time as "a figment of people's romantic imagination, – a wild meadow threaded with rusty nails high above the street – that was visited primarily by adventuresome truants and graffiti artists". In the context of the 1990s urban renewal in the surrounding neighbourhoods, in particular in the Meatpacking District and Chelsea, the ideas about the High Line changed, and the not-for-profit Friends of the Highline, funded by two local residents, increasingly gained traction in their quest for its preservation and reuse as an elevated park. The interest in developing the High Line as an elevated park, an urban promenade, aligned with the agendas pushing further urban renewal and gentrification in the surrounding neighbourhoods, and the demographics involved in this urban transformation became quickly involved: *New York Times* architectural critic Nicolai Ouroussof wrote in 2006 " . . . the High Line quickly became the favorite cause of New York's rich and powerful".[24] After a number of complex real-estate deals, swapping development rights and air rights, the securing of financing through public–private partnerships, New York firms Field Operations (landscape architecture) and Diller Scofidio+Renfro (architecture) won an open-design competition.[25]

Joel Sternfeld's photographs of the High Line, presenting it as an industrial ruin and its emergent ecologies as a form of urban wilderness and invoking ideas and visual experiences of the sublime (see Figure 5), were hugely successful in branding the Highline and garnering public support.[26] His images (much reminiscent of eighteenth- and nineteenth-century landscape paintings) operate within and employ traditions of the picturesque, framing and rendering visible a post-industrial urban sublime and aestheticising a form of urban nature where non-human processes insert and assert themselves into the margins and neglected interstices of the city. Interestingly, as much as Sternfeld's photos render a "perfect ruin" in the picturesque tradition, they also succeed in framing and revealing the representational agency of the High Line itself before the design was implemented: the ongoing processes of growth and decay, manifested through successional vegetation and emergent ecologies form the most transgressive elements, visualising the threat to the systems of social and spatial order operating in and around the sites.

Field Operations' and Diller Scofidio+Renfro's design for the High Line is deeply rooted in these traditions of the picturesque. It is noteworthy that the High Line's design is distinctively at odds with Field Operation's principal James Corner's anti-picturesque ideas about landscape and its agency . His writings have been adamant in arguing for the primacy of process and performance over form, and dismissed the picturesque as an outdated, reductive and inappropriate approach to the design and understanding of landscape.[27] The design removed the existing emergent and transgressive ecologies with their "wild" appearance, and replaced them with carefully designed and choreographed artificial ecologies, complimented by the removal of the gritty appearance of an industrial ruin, replacing industrial materials by slick, sophisticated design elements with refined materials and finishes. It disappointed expectations of retaining at least some of the High Line's existing vegetation and emergent ecologies, upsetting Sternfeld, who opposed the removal and replacement of what he considered the primary qualities of the High Line (see Cataldi *et al.* 2011).

While the "before" shows much more qualities of an unmaintained, "gritty", "wild" and "forbidden" place with a much clearer expression of non-human processes, the "after" appears much more like a garden, the non-human processes tamed, choreographed and carefully controlled and managed to create a preconceived elaborate and "sleek" aesthetic experience emphasising the visual (see Figure 5).

The selected plant material[28] is controlled by maintenance regimes that are not just exclusive and restrictive, but also extensive: the maintenance budget for the High Line exceeds by far, at about $50,000 per acre and year, the $1000 per acre and year average for parks in New York City. Plant species are arranged in ecotypes – many of them have

Figure 5.   Constructing the picturesque: before – after. Note the use of Sternfeld's photograph on the billboard, as seen from the High Line.
Source: Photographs by the author.

little to do with any imaginable outcome of successional processes on the High Line. Their obvious, overmaintained artifice is a distinct nod to the picturesque – providing a romantic, domesticated and aestheticised version of urban nature or the "urban wilderness" – a vision and version of urban nature that is predominantly visual in nature.

Most picturesque works of landscape architecture produce explicit and implicit narratives[29]: The High Line as a promenade moves the visitor from woodlands to grasslands, to trees and shrubs, some of them ornamental, to grasslands, back to an area of denser woodland vegetation, and grasslands at phase 2's northern terminus on 30th street. The experience of different ecotypes in succession suggests a narrative of pristine pastoral nature, excluding the surrounding urban conditions, replacing the possibility of a critical engagement with the processes and the parameters of its cultural production in an urban context with notions of escapism (Tuan 1998).

Contrary to Duisburg-Nord, the High Line provides a highly controlled, choreographed and pretexted experience, foregrounding the visual experience, much more akin to a promenade with a sequence of tableaus designed to be seen, but not otherwise interacted with. The planting design as well as the placement of benches, viewing platforms, etc. enforces a particular experience and ocular regime, with little possibility for deviation. Behaviour is tightly controlled, appropriate behaviour regularly enforced by other visitors and the ubiquitous presence of maintenance staff and park police.

The exclusion of visible manifestations of non-human processes[30] makes impossible meaningful and formative experiences of a more authentic type of urban nature that could be considered challenging or resistant to the hegemonial agendas of urban renewal – by constructing and maintaining the type of romantic nature and its inherent qualities as exclusive aesthetic object, and by controlling any type of experience and interaction (see Figure 6). Aesthetic experience is reduced to the mere act of looking, any experience that is not merely visual limited to passive reception is explicitly discouraged and policed accordingly.

The replacement of the ecologically transgressive corresponds to the exclusion and displacement of the socially transgressive (paralleled by the "clean-up" of the Meatpacking

Figure 6.   Highline experience.
Source: Photograph by the author.

District with its pasts of marginal industrial production and subsequent image of sexual deviancy). Urban renewal and gentrification processes were already well on their way and displaced large numbers of the poorer residents before the High Line, but it will without doubt be instrumental in accelerating these processes (Cataldi *et al.* 2011, p. 377).

Traditional picturesque aesthetics and ideology emphasise the visual and are ultimately exercises in the production of images with distinct and hidden agendas, and the High Line's aesthetic performance operates within this framework. The rebranding of the Lower West Side relies on the production of images (of city, urbanity, nature, culture, etc.), and the High Line operates simultaneously as visual aesthetic object and as a "viewing platform", a device that facilitates privileged views of the city. Traditional picturesque ideology has deployed its aesthetics to successfully hide the actual authorship of space and succeeded in "naturalising" hegemonial power. The High Line operates along similar lines – by its exclusion of transgressive human and non-human processes and their physical manifestations, it replaces the possibility of encountering a more authentic and inclusive version and vision of urban nature with its reductive, domesticated artefact – an artefact that cements and expands existing power differentials, and excludes ideas about the relationship between human and non-human processes that might challenge the hegemony of global capital flows in imagining alternative versions of urbanity and urban development.

Images themselves tend to be static, producing an appearance that is at odds with the experience of dynamic change. Much in the picturesque tradition, images privilege particular spatial conditions (the perspectival construction of imagery privileges one particular point of view, and one particular direction of the gaze), but they also pick one particular moment in time, thus limiting the ability to imagine any kind of change. The High Line, as it is a translation of such imagery, clearly suggests a static imagination of urban nature – a domesticated version that is predictable and benevolent, subtly invoking ideas about non-human systems as homeostatic, balanced and ultimately "sustainable" (Turner 1994, pp. 39–42) – the sustainability achieved without challenging capitalist-hegemonial systems and their obvious production of environmental inequality.

The analysis of photographic imagery often focuses on the act and implications of "framing", beginning with an interrogation of what is included and excluded: analogically, the High Line can be interpreted as an act of framing and producing a distinct type of urban nature. The following typology is an expansion of concepts of nature originating in European antiquity and developed through the eighteenth-century picturesque and twentieth-century urban ecology (see Hard 1995, Hunt 2000, Le Roy 2002):

1st nature = natura naturans = wilderness, uninfluenced by humans

2nd nature = natura naturata = agriculture

3rd nature = the garden, containing idealised versions of 1st and 2nd nature

4th nature = 1st nature re – inhabiting 2nd and 3rd nature = transgressive and emergent ecologies.

The High Line then frames urban nature as 3rd nature, constructing a traditional picturesque *garden*, excluding the emergent and transgressive ecologies, and the non-human processes they are produced by and signify. Its actual performance is relegated to aesthetic appearance. As such, the High Line's primary agency lies in the *performance of (ecological) aesthetics, rather than an aesthetics of (ecological) performance*, rendering it more as a "spectacle" (Debord 1967) or as "hyperreality" (Baudrillard 1981), in which the signifier (High Line's domesticated version of urban nature) replaces the signified (nature and non-human processes, expressed as emergent ecologies).

The "new appropriation" of the picturesque as ideological and representational tool in the capitalist-hegemonial control of emergent constructs of ecology, urban nature and urbanity, serving the stabilisation and control of uneven developmental patterns emphasises the shifting role of landscape in this context: Cosgrove (1998, p. XIV) argues that " ... landscape constitutes a discourse, through which identifiable social groups historically have framed themselves and their relations with both the land and with other human groups, and that this discourse is closely related epistemically and technically to ways of seeing .... ." The discourse on replacing ("domesticating") the transgressive elements of urban nature on the High Line is played out through the production of imagery and actual (aesthetic) experience of landscape, reducing it to its representational content and semiotic function as a signifier (Corner 1992, 1999, Cosgrove 1998).

While the High Line, much like any landscape, is invariably a simultaneous ruin of nature and culture, its emphasis is clearly on the processes of human cultivation, producing an artifice that involves non-human processes, forms and materials, but frames and controls them tightly. While Duisburg-Nord reveals the agency and workings of human and non-human processes, the High Line camouflages the human hand and the fact that it is mostly a ruin of nature and edifice of culture.

Conclusion

The relevance and agency of Duisburg-Nord and the High Line, 20 years apart, lies very much in their function as representations of ideas about nature, culture, sustainability and urbanity – their iconic appearance has, along with their predecessor of Gasworks Park in Seattle (1971), informed and shaped practices and discourses in their particular cultural contexts. As a progression these projects highlight changing rhetorics on the post-industrial city and the role of non-human processes: from Gasworks Park – an exercise in the classic picturesque, to Duisburg-Nord – primarily concerned with the interplay of human and non-human processes in an open-ended framework, to the Highline – which seemingly reverts to a classic-picturesque approach.

Additionally, these illustrate the complex, opaque, enmeshed and entangled relationship between ecological and aesthetic values. This relationship in general can be one of mutual dependency, and often fraught with fundamental misunderstandings, some accidental, some intentional. Constructed ecologies involve the translation of ecological values into landscape architectural or architectural form and necessarily have an aesthetic/ picturesque dimension (Meyer 2008). More insidiously, aesthetic preferences are often unintentionally translated into and argued as ecological values (Botkin 1990, Nassauer 1995, Langhorst and Kambic 2009). Correspondingly, the aesthetics of existing emergent and transgressive ecologies have also been used to argue particular ecological values.

The label "transgressive" clearly alludes to a marginalisation of such emergent ecologies (or, of the non-human processes they represent and are a result of, frequently called "nature"). Lefebvre, Hetherington, Massey, Soja and others all locate spaces of resistance in the margins. Post-industrial sites are marginalised, and replete with margins that emerge when the predominant ordering systems and their maintenance and use regimes weaken or cease to operate.

Post-industrial sites operate as heterotopias[31] and need to continue to do so if they are to be successful as locations of the renegotiation of human–nature relationships in urban contexts. This renegotiation is posited as an ongoing process of contestation between multiple human and non-human processes, and counters traditional concepts of ecologies (or "nature") as homeostasis or "ideal states that are perfect and should not be tampered with" (Turner 1994, p. 39). Such "ideal states", as frequently argued in the context of sustainable development, are suspect of "naturalising" existing political and economic systems, using ecological values. On most post-industrial sites the (cultural/capitalist) ordering during their active phases as sites of production is eventually subverted by non-human processes and their spatial expressions (as emergent and transgressive ecologies). The different treatments of these ecologies on post-industrial sites consequentially reveal vastly different underlying constructs of social order.

While Duisburg-Nord engages emergent ecologies as an exploratory, open-ended and fluid interaction between human intervention and non-human process, the High Line replaces actual successional ecologies with an artifice based on picturesque traditions. These different representations and aesthetications of non-human processes construe both sites as location, icon and instrument of deterministic-hegemonial and transgressive-marginal agendas, emphasising their role in postmodern and post-colonial discourses on the nature of urbanity and sustainable urban renewal – discourses central to the concepts of "landscape urbanism" and "ecological urbanism".

In this conflict it seems critical to preserve the heterotopian qualities that characterise "authentic" urban nature and make it both attractive and necessary: the presence and experience of alternate and competing ordering system, that allow to engage the whole gamut of different processes and their physical manifestations as an ongoing practice.

Three concerns should guide the discourses and practices relating to the roles of post-industrial sites in general and emergent ecologies in particular, carefully identifying the processes and agents (human and non-human) involved in the production of urban natures and ecologies:

(1) *The instrumentality and agency of emergent ecologies.* Aside from the performance of ecological processes and functions, emergent ecologies have representational and aesthetic agency and can act as signifiers for concepts of social order and of culture–nature relationships. *Aesthetics of performance* (as in Duisburg-Nord) are opposed to the *performance of aesthetics* (as in the High Line), the latter reducing the complex agency of landscape and urban nature to that of a compelling visual narrative supporting hegemonial-capitalist conditions and agendas.

(2) *The type of aesthetic experience determines the understanding and conceptualisation of urban nature(s).* Aesthetics as complex, situated, participatory and multi-sensorial experience (Berleant 1992, references environmental aesthetics as "being in process with the environment") produces different understandings, interpretations and concepts of self and urban natures than aesthetics as a tightly controlled, predominantly visual experience that reduces urban natures to an aesthetic object that is passively perceived. Assuming a close connection between the experience of urban natures and ecologies and the formulation of identity, the former empowers and enables an active participation in the production of urbanity, while the latter disables all potential for involvement, marginalising already marginalised people even more.

(3) *The type of urban nature that is produced and experienced is reflective of underlying constructs of political, economic and social ordering systems, and has different agency in supporting or subverting hegemonial-capitalist constructs of the post-industrial city.* Fourth nature, as a productive heterotopian concept, is inclusive and empowers forces and agents traditionally marginalised in the production of urbanity, distinctively involving non-human processes that are not tightly controlled and managed. Third nature is based on high levels of control, and

excludes/disempowers non-human processes as well as marginalised people from the production of urban nature. The resulting authoritative expression of nature idealises a particular construct of nature–culture relationships that is homeostatic and stabilises existing systems of social order, ultimately rendering it more of a theme park. People's agency is limited to that of passive consumption of a ready-made product, a product whose authorship is hidden and naturalised.

Strategies that aestheticise authentic urban natures by replacing them with domesticated images of a picturesque nature then have to be considered insidious, ethically suspect and inherently unjust, as they deprive already marginalised communities from

(a) an inclusive aesthetic experience of authentic urban nature, and
(b) an authentic expression of non-human process by displacing transgressive ecologies
(c) heterotopian conditions as spaces of empowerment, due to their inherent elimination, and
(d) any meaningful participation in inclusive and open-ended discourses on urban transformation, as they are limited in scope by the hidden authoritative construction of possible and existent urban natures.

Aesthetic justice[32] and the ability to experience different urban natures then becomes a critical, necessary and, so far, neglected part of the discourses on the right to the city. Emergent ecologies need to be considered not just in their function as habitat and provider of other infrastructural functions, but as complex instruments with multiple agencies within the highly contested broader trajectories of urbanisation and sustainable development.

James Corner, co-designer of the High Line and a prominent participant in the discourses on urbanity and the nature of Cities, suggests

> landscape (...) as a metaphor for inclusive multiplicity and pluralism, as in a kind of synthetic "overview" that enables differences to play themselves out. (...) In these terms, landscape may still embrace naturalistic and phenomenological experience but its full efficacy is extended to that of a synthetic and strategic art form, one that aligns diverse and competing forces (social constituencies, political desires, ecological processes, program demands, etc., into newly liberating and interactive alliances. (Corner 1999, pp. 2–4)

This emphasis on landscape as a central agent of cultural change, and the instrumentalisation of emergent ecologies in particular, involve different conceptualisations of "natural", without making explicit the wide range of ideas about "nature" they reference:

> Given the many meanings and definitions of what is natural, appeal to nature as authority for human actions is problematic. Any approach to landscape design based on the notion that nature is singular or its meaning universal or eternal is sure to founder. The emphasis should be on a spirit of inquiry and exploration rather than close-minded certainty. Emotional and rhetoric and doctrinaire positions will not advance this agenda, but rather a more reasoned, self-critical, inclusive approach which acknowledges the plurality of human values and motives embedded in ideas of nature and authority. (Spirn 1997, pp. 259–260)

As places that both enable and necessitate an open-ended exploration of urban nature as an ongoing negotiation of human and non-human processes, and as sites that perform as both passive ground and active agent in the constant and continuous remaking of place, post-industrial sites should be positioned at the centre of the discourses on the nature of cities, urbanisation and sustainable development.

The current discourses on "sustainable" and "resilient" cities tend to ignore important questions about the "contradictions of capital accumulation and circulation, about uneven development, about enabling political structures, about state strategies of 'growth machine' branding" (Slater 2014), and are silent about its most insidious process: the "accumulation by dispossession" (Harvey 2004, 2005, Staeheli *et al.* 2002) or, in other words, the displacement of communities for capital accumulation. This dispossession is aided by the displacement of emergent and transgressive ecologies as expressions and their replacement with the spectacles of constructed urban natures in "post-industrial theme parks". This paper contends that post-industrial sites, if designed with a clear understanding of their heterotopian character and construction as simultaneous ruins of culture and nature, can serve as Soja's (1996) "third-space" – as a place in the margins that can serve as a location of resistance against the dominant neoliberal agendas of urban growth camouflaged as sustainability and resilience – by providing empowering, inclusive and alternative versions and experiences of "urban nature" that models socially, environmentally and aesthetically just cities. Future research shall explore how and to what degree Landschaftspark Duisburg-Nord, the High Line and countless other current and future projects succeeding them, will impact these discourses as well as the resultant and attendant physical, social, cultural and ecological urban realities.

Notes

1. The author has visited Landschaftspark Duisburg-Nord at least twice annually since 1985, and the Highline multiple times since 2006. He visited and experienced both sites multiple times before the design proposals were implemented.
2. "Sublime" is a critical concept within picturesque, aesthetics, ideology and style (see Knight [1805] 1975, Burke [1757] 1990, Meyer 1998, Herrington 2006).
3. For an overview of the fallacies and implications of this approach, see e.g. Oelschlager (1991), Cronon (1996), Corner (1997), Turner (1994) and Casey (2009). To avoid the normative connotations of the traditional terms "nature" and "culture", the terms "non-human processes" and "human processes" are used when necessary.
4. For an overview of the history and application of these concepts, see Corner (1999), Waldheim (2006) and Mostafavi (2010).
5. Environmental justice in contemporary discourses emphasises questions of environmental quality (such as air and water quality). I argue that the quality of the spaces inhabited by humans, both build and "natural" are an equally important component of environment, as they facilitate, mitigate, enable or discourage possibilities of interaction and inhabitation. This extended understanding of environmental justice aligns with Lefebvre's, Harvey's and Mitchell's concept of the "right to the city" as a right to participate in the physical, conceptual and ideal spaces that make up a city.
6. "Authentic" is obviously a highly contested concept in itself. The Oxford English Dictionary defines authentic as "*real, actual, genuine, original, first-hand, really proceeding from its stated source*" (OED, emphasis by author). In this sense, emergent and transgressive ecologies are indeed an unmitigated and unmanipulated result of non-human processes responding to the particular conditions of a site at a particular point in time. They are not "edited" by any human intervention.
7. This applies DeCerteau's (1984) concept of spatial tactics, as they are employed by marginalised cultures against the spatial strategies of hegemonial elites in control of the mechanisms of spatial production.
8. The concept of "transgressive ecologies" proposed here is developed as an analog to the framework of culturally transgressive behaviours laid out in Soja's "Thirdspace" (1996).
9. Recent literature denotes an interest in the functional and performative aspects of ecosystems and landscapes, as denoted by discussions on "landscape as infrastructure" (Raxworthy and Blood 2005, Poole 2005a, 2005b), "landscape performances" and "ecosystem services" (*Landscape and Urban Planning*, 109 (2013), Special issue on "Ecosystem Services").

10. Mitchell (1994, p. 14) contends that

> landscape itself is a physical and multisensory medium (earth, stone, vegetation, water, sky, sound and silence, light and darkness etc.) in which cultural meanings and values are encoded, whether they are *put* there by physical transformation of a place in landscape gardening or architecture, or in a place formed, as we say, by nature.

Knowledge of the particular concepts and traditions within landscape architecture and related disciplines and professions can be considered privileged, as it is usually limited to members of these and related fields. "Contingent upon the thematic dimension of Picturesque works is the notion that their interpretation depends upon an educated reader" (Mitchell 1994, p. 14).

11. It is important to acknowledge that Landschaftspark Duisburg-Nord and the High Line are different in terms of immediate spatial contexts, scale, complexity, length of abandonment and environmental hazards. They are similar in the ways they are instrumentalised in discourses on urban renewal and the conversion of neighbourhoods, in particular in how their aesthetic appearance and performance utilised to represent obvious and hidden agendas with their respective ideas on the nature of the city.

12. The resultant "pastoral" landscapes with their signature style (curving paths winding through undulating hills of grass punctuated by groupings of deciduous trees continued to influence the formal conventions of public parks and private estates until the twentieth century (Herrington 2006).

13. Herrington (2006, p. 23) discerns three faces of the picturesque: picturesque style, picturesque ideology and picturesque aesthetics.

14. Iconic works of landscape architecture in the picturesque style either incorporated existing ruins (e.g. the Cistercian Abbeys at Rielvaux Terrace and Fountain's Abbey in Yorkshire, UK) or involved the building of artificial ruins, so-called follies to create the desired aesthetic experiences and interpretations (as in the picturesque parks of, e.g. Rousham, Stourhead, Stowe in the UK) (Hunt 2000).

15. A decision of the IBA steering committee, based on it being representative of the ongoing processes and consequences of the deindustrialisation of the region, on its iconic appearance, the urgency of addressing contamination in close proximity to residential areas, and, last but not least, on the iconic character of the industrial ruins. Most IBA publications involved professional photographs of Duisburg-Nord. It could in fact be argued that the iconic imagery of Duisburg-Nord played as important a role in garnering public support for the project as Sternfeld's photographs did for the High Line. It is important to qualify here that the persuasive agency of the Duisburg-Nord photographs did not extend to the neighbourhoods and communities surrounding the site, nor did they affect former workers in similar industries in the regions.

16. About US$2 billion.

17. An approach related to picturesque traditions of landscape architecture. For a critical discussion, see Herrington (2006), and the discussion of picturesque ideology in the context of the High Line below.

18. Hargreaves describes this as "attenuated wilderness" (2007, p. 168). See also Nassauer 1995.

19. The concept has been used widely in urban studies, critical and cultural geography, cultural studies, etc. Originally introduced by Foucault and Lefebvre, developed further by Harvey and Soja, excellent overviews are provided by Massey (2005) and Hetherington (1997). The classic definition of heterotopias as sites of competing or alternate ordering might have been written for post-industrial sites – the distinct (cultural/capitalist) ordering during their active phases as sites of production is subverted by non-human processes and their spatial expressions. Because of the distinct visibility, and appearance of these alternate ordering systems, post-industrial sites, at least in their undersigned state, attain a legibility of heterotopian conditions that can be considered unique.

20. James Corner's assertion is distinctively at odds with the approach and conceptual understandings of landscape and ecology involved in his design for the High Line. His ideas here reference, among others, Haraway (1991) and Latour (1993). See also note 30.

21. Turner (1994) annotates this contention as an "unpacking of the central term 'sustainable'" (p. 39). "Thus, the idea of 'sustainability' and general homeostasis is a profoundly unnatural goal" (p. 41). "Nature is not and has never been static and unchanging" (p. 40), " ... it is a system to subvert sustainability and maintainability" (p. 41).

22. Such activities involve the consumption of alcohol, drugs and other illegal substances, as well as a wide range of sexual acts. The site qualities that discourage or disable surveillance are of great importance here. More organised activities, such as spontaneous musical performances, secret and illegal raves, etc. have found their designed accommodations on the site, but frequently occur in other areas.

23. DeCerteau (1984) discerns between the passive visual reception of urban space (the voyeur) and the active participation in constructing urban space through acts of use (he emphasises walking), the latter amounting to the "writing" of urban space.

24. For an excellent overview over the process of securing and building the High Line, see Cataldi *et al.* (2011).

25. Both firms have a reputation for being on the cutting edge of design in urban conditions, both firms can be considered academic practices, as their principals hold faculty appointments at prestigious design schools known for the critical and innovative positions.

26. This recalls the role of paintings, etchings and photography in the rendering of "wilderness" as a critical part of the early national park movement in the USA.

27. James Corner is the principal of Field Operations, one of the few contemporary writers on landscape architectural theory and one of the progenitors of the field of landscape urbanism (Corner 1990, 1991, 1992, 1997, 1999).

28. The planting design was developed by Piet Oudolf, one of the world's most renowned planting designers. The planting palette involved 210 species, 40% occurred on the High Line in its abandoned stage. The introduced plants were chosen for their appearance (texture, foliage and bloom), and for their ability to withstand the challenging soil and microclimatic conditions on the highline. A significant percentage of those plants are not native to the region (Martin 2009).

29. Cronon (1992) provides an account of the problems inherent in environmental narratives. By extension, narratives that are environmentalised are equally or even more problematic.

30. It could of course be argued that the planting design itself provides the experience of non-human process, but these processes are so tightly controlled and subsumed under an expected appearance that the human element is dominant to the point of being exclusive.

31. See Hetherington (1997, p. 41) for a list of applications of the concept of "heterotopia".

32. The concept of aesthetic justice has been argued extensively by Beardsley (1982). For a critique in the context of spatial design and urban planning, see Mattila (2002).

References

Arendt, H., 1958. *The human condition*. Chicago: Chicago University Press.

Baudrillard, J., 1981. *Simulacres et Simulation*. Paris: Éditions Galilée.

Beardsley, M.C., 1982. Aesthetic welfare, aesthetic justice and educational policy. *In*: M.J. Wreen and D.M. Callen, eds. *The aesthetic point of view: selected essays: Monroe C. Beardsley*. London: Cornell University Press, 111–124.

Berleant, A., 1992. *The aesthetics of environment*. Philadelphia, PA: Temple University Press.

Berrizbeitia, A., 2001. Horizons of undecidability. *In*: J. Czerniak, ed. *Case: Downsview Park*. New York: Prestel, 117–125.

Botkin, D.B., 1990. *Discordant harmonies: a new ecology for the 21st century*. Oxford: Oxford University Press.

Burke, E., 1990. *A philosophical enquiry into the origin of our ideas of the sublime and the beautiful*. Oxford: Oxford University Press (original published in 1757).

Casey, E., 2009. Going wild in the land. *In*: E. Casey, ed. *Getting back into place: toward a new understanding of the place-world*. Bloomington: Indiana University Press, 227–268.

Cataldi, M., *et al.*, 2011. Residues of a dream world: the high line, 2011. *Theory, Culture, Society*, 28 (7–8), 358–389.

Corner, J., 1990. A discourse on theory I: sounding the depths. Origins, theory and representation. *Landscape Journal*, 9 (2), 61–78.

Corner, J., 1991. A discourse on theory II: three tyrannies of contemporary theory and the alternative of hermeneutics. *Landscape Journal*, 10 (2), 115–133.

Corner, J., 1992. Representation and landscape: drawing and making in the landscape medium. *Word and Image*, 8 (3), 243–275.

Corner, J., 1997. Ecology and landscape as agents of creativity. *In*: G. Thompson and F. Steiner, eds. *Ecological design and planning*. New York: John Wiley, 81–108.

Corner, J., 1999. Recovering landscape as cultural practice. *In*: J. Corner, ed. *Recovering landscape: essays in contemporary landscape architecture*. New York: Princeton Architectural Press, 1–25.

Cosgrove, D., 1998. *Social formation and symbolic landscape*. Madison: University of Wisconsin Press.

Creswell, T., 2004. *Place. A short introduction*. Oxford: Blackwell.

Cronon, W., 1992. A place for stories: nature, history and narrative. *The Journal of American History*, 78 (4), 1347–1376.

Cronon, W., 1996. The trouble with wilderness; or, getting back to the wrong nature. *In*: W. Cronon, ed. *Uncommon ground: rethinking the human place in nature*. New York: W. Norton, 69–90.

Debord, G., 1967. *La société du spectacle*. Paris: éditions Buchet-Chastel.

DeCerteau, M., 1984. Walking in the city. *In*: M. DeCerteau, ed. *The practice of everyday life*. Berkeley: University of California, 91–110.

Edensor, T., 2005. *Industrial ruins: spaces, aesthetics and materiality*. London: Berg.

Evernden, N., 1992. *The social creation of nature*. Baltimore, MD: Johns Hopkins University Press.

Foote, K., 2003. *Shadowed ground: America's landscapes of violence and tragedy*. Austin: University of Texas Press.

Foster, J., 2010. Off track, in nature: constructing ecology on old rail lines in Paris and New York. *Nature and Culture*, 5 (3), 316–337.

Haraway, D., 1991. *Simians, cyborgs and women: the reinvention of nature*. New York: Routledge.

Hard, G., 1995. *Spuren und Spurenleser: Zur Theorie und Ästhetik des Spurenlesens in der Vegetation und anderswo*. Osnabrück: Universitätsverlag Rasch.

Hargreaves, G., 2007. Large parks: a designer's perspective. *In*: J. Czerniak and G. Hargreaves, eds. *Large parks*. New York: Princeton Architectural Press, 121–171.

Harvey, D., 1973. *Social justice and the city*. London: Edward Arnold.

Harvey, D., 1996. *Justice, nature, and the geography of difference*. London: Blackwell.

Harvey, D., 2004. The "new" imperialism: accumulation by dispossession. *Socialist Register*, 40, 63–87. Available from: http://socialiststregister.com [Accessed 14 January 2012].

Harvey, D., 2005. *A brief history of neoliberalism*. Oxford: Oxford University Press.

Herrington, S., 2006. Framed again: the picturesque aesthetics of contemporary landscapes. *Landscape Journal*, 25 (1), 22–37.

Hester, R., 2006. *Design for ecological democracy*. Boston: MIT Press.

Hetherington, K., 1997. *Badlands of modernity: heterotopia and social ordering*. London: Routledge.

Hunt, J. and Willis, P., eds., 1975. *The genius of the place: the English landscape garden, 1620–1820*. New York: Harper & Row.

Hunt, J., 2000. *Greater perfections: the practice of garden theory*. Philadelphia: University of Pennsylvania Press.

Hunt, J. and Willis, P., eds., 2002. *Tradition and innovation in French garden art*. Philadelphia: University of Pennsylvania Press.

Huyssen, A., 2003. *Present pasts: urban palimpsests and the politics of memory*. Palo Alto, CA: Stanford University Press.

Ingold, T., 2000. The temporality of landscape. *In*: T. Ingold, ed. *The perception of the environment. Essays on livelihood, dwelling and skill*. London: Routledge, 189–208.

Iovine, J., 2009. All aboard the high line. *The Wall Street Journal*, 23 June, p. D7.

Kaika, M., 2009. Landscapes of energy: hydro-power from techno-natures to retro-natures. *Harvard New Geographies*, 2, 102–110.

Knight, R., 1975. An analytical inquiry into the principles of taste. *In*: J. Hunt and P. Willis, eds. *The genius of place: the English landscape garden 1620–1820*. New York: Harper & Row, 248–250 (original published in 1805).

Langhorst, J., 2004. Re-covering landscapes: derelict and abandoned sites as contested terrains. *ICON: The Journal of the International Committee for the History of Technology*, 10, 65–80.

Langhorst, J., 2012. Recovering place – on the agency of post-disaster landscapes. Special Issue "Post-Disaster Landscapes", *Landscape Review* 14 (2), 48–74.

Langhorst, J. and Kambic, K., 2009. Massive change, required: nine axioms on the future of landscape (architecture). *KERB – Journal of Landscape Architecture*, Royal Melbourne Institute of Technology, #17 Is landscape architecture dead? 105–110.

Latour, B., 1993. *We have never been modern*. Cambridge, MA: Harvard University Press.

Lefebvre, H., 1996. Right to the city. *In*: E. Kofman and E. Lebas, trans. and eds. *Writings on cities*. Oxford: Basil Blackwell, 63–177.

Le Roy, L.G., 2002. *Natuur, cultuur, fusie*. Amsterdam: Nederlands Architectuur Instituut.

Louv, R., 2005. *Last child in the woods*. Chapel Hill, NC: Algonquin Books.

Marot, S., 2003. *Sub-Urbanism and the Art of Memory*. London: Architectural Association.

Martin, G., 2009. New York's hanging gardens. *The Guardian, Observer Magazine*, 8 Nov.

Massey, D., 2005. *For space*. London: Sage.

Mattila, H., 2002. Aesthetic justice and urban planning: who ought to have the right to design cities? *GeoJournal, Social Transformation, Citizenship, and the Right to the City*, 58 (2/3), 131–138.

Meyer, E., 1998. Seized by sublime sentiments: between Terra Firma and Terra Incognita. *In*: W. Saunders, ed. *Richard Haag: Bloedel Reserve and Gas Works Park*. New York: Princeton Architectural Press, 5–28.

Meyer, E., 2008. Sustaining beauty. The performance of appearance. A manifesto in three parts. *Journal of Landscape Architecture (JoLa)*, Spring 2008, 6–23.

Mitchell, W., ed. 1994. *Landscape and power*. Chicago: University of Chicago Press.

Mitchell, D., 2003. *Right to the city: social justice and the fight for public space*. New York: Guilford.

Mostafavi, M., 2010. Why ecological urbanism? Why now? *In*: M. Mostafavi and G. Doherty, eds. *Ecological urbanism*. Boston, MA: Harvard University Graduate School of Design, Lars Müller, 12–53.

Nassauer, J., 1995. Messy ecosystems, orderly frames. *Landscape Journal*, 14 (2), 161–170.

Nora, P., 1989. Between memory and history: Les Lieux des Memoire. *Representations, Special Issue: Memory and Counter-Memory*, 26, 7–24.

Oelschlager, M., 1991. *The idea of wilderness: from prehistory to the age of ecology*. New Haven, CT: Yale University Press.

Poole, K., 2005a. Potentials of landscape as infrastructure part 1: six and a half degrees of infrastructure. *In*: J. Raxworthy and J. Blood, eds. *The MESH book: infrastructure/landscape*. Melbourne: RMIT Press, 18–49.

Poole, K., 2005b. Potentials of landscape as infrastructure part 2: creative infrastructures – dynamic convergences of ecology, infrastructure and civic life. *In*: J. Raxworthy and J. Blood, eds. *The MESH book: infrastructure/landscape*. Melbourne: RMIT Press, 180–199.

Raxworthy, J. and Blood, J., eds. 2005. *The MESH book: infrastructure/landscape*. Melbourne: RMIT Press.

Robinson, S., 1988. The picturesque: sinister dishevelment. *Threshold*, IV, 76–81.

Schuhmacher, P. and Rogner, C., 2001. After Ford. *In*: G. Daskalakis, C. Waldheim, and J. Young, eds. *Stalking Detroit*. Barcelona: Actar, 48–56.

Slater, T., 2014. *The resilience of neoliberal urbanism* [online]. Available from: http://www.opendemocracy.net/opensecurity/tom-slater/resilience-of-neoliberal-urbanism [Accessed 3 February 2014].

Smith, N., 1984. *Uneven development: nature, capital and the production of space*. Oxford: Blackwell.

Soja, E., 1996. *Thirdspace*. Malden, MA: Blackwell.

Soule, M. and Lease, G., eds. 1995. *Reinventing nature? Responses to postmodern deconstructionism*. Washington, DC: Island Press.

Spirn, A., 1997. The authority of nature. *In*: J. Wolschke-Bulmahn, ed. *Nature and ideology: natural garden design in the 21st century*. Washington, DC: Dumbarton Oaks, 249–261.

Staeheli, L., Mitchell, D., and Gibson, K., 2002. Conflicting rights to the city in New York's community gardens. *GeoJournal, Social Transformation, Citizenship, and the Right to the City*, 58 (2/3), 197–205.

Swyngedouw, E., 2010. Trouble with nature: ecology as the new opium for the people. *In*: J. Hillier and P. Healey, eds. *Conceptual challenges for planning theory*. Farnham: Ashgate, 299–320.

Talen, E., 2010. *A tire in the park* [online]. Available from: http://bettercities.net/news-opinion/blogs/emily-talen/13579/tire-park [Accessed 14 January 2012].

Tilley, C., 1994. *A phenomenology of landscape. Places, paths and monuments*. Oxford: Berg.

Tuan, Y., 1974. *Topophilia: a study of environmental perception, attitudes, and values*. Englewood Cliffs, NJ: Prentice-Hall.

Tuan, Y., 1998. *Escapism*. Baltimore, MD: Johns Hopkins University Press.

Turner, F., 1994. The invented landscape. *In*: D. Baldwin, J. De Luce and C. Pletsch, eds. *Beyond preservation: restoring and inventing landscapes*. Minneapolis: University of Minnesota Press, 35–64.

Waldheim, C., 2006. Landscape as urbanism. *In*: C. Waldheim, ed. *The landscape urbanism reader*. New York: Princeton Architectural Press, 35–53.

Weilacher, U., 2008. *Syntax of landscape: the landscape architecture of Peter Latz and partners*. Basel: Birkhauser, 102–133.

Index